DEALING WITH RISK

Howard Margolis

To Chet Cooper

CONTENTS

ACKNOWLEDGMENTS

Principal support for this study has come from the Global Studies Program at Battelle Pacific Northwest Laboratory, which proved to provide intellectual as well as material sustenance. So it is a pleasure to thank Gerry Stokes, Bill Pennell, and John Clarke of that program. An old friend and colleague, Chet Cooper, was midwife to this effort, and I owe him a large debt for this and also for past moral support. The intellectual capital I was able to draw on from earlier work was substantially made possible by the National Science Foundation. In particular, chapter 3 is based on work sponsored by the NSF Decision Research and Management Science program. As a reader will find, ideas from that chapter are drawn on throughout the study. Much of the work on cognitive illusions was done during an extended visit to the Australian National University in Canberra. I am indebted to Geoffrey Brennan and the Research School of the Social Sciences for this stimulating opportunity. And a final gap was filled (with the support of Bill Niskanen and Aaron Wildavsky) by a grant from the Earhart Foundation. I am grateful to all. Of course, the views argued here are solely my own.

INTRODUCTION

As this is written, the prevailing account of expert versus lay conflicts of risk intuition on such matters as nuclear waste and pesticides is that experts focus on a very narrow range of consequences, but ordinary people have a much richer sense of what is involved in choices about risk. So experts may feel comfortable with a level of precautions that seems wholly inadequate to ordinary people.

And experts indeed typically assess subtle risks—risks of the sort that are the focus of expert/lay controversy—almost exclusively in terms of quantitative assessment of injuries or illnesses or fatalities. But the usual claim has been that ordinary citizens see things differently because they are concerned about additional dimensions of risk such as voluntariness, and about trust in authorities responsible for managing the risk, and about risk to future generations. An exemplar of this view that I use a good deal itemizes, in fact, 19 such dimensions.

The result of the analysis here is to turn the usual story inside out. Responses on the extra dimensions, I am led to argue, do not plausibly *explain* why lay perceptions of risks conflict with expert assessments when they do (in fact, much of the time they do not, which is an important point that has to be accommodated). Rather, I try to show, there are good reasons to believe that the responses along the many extra dimensions more plausibly reflect perceptions of risk than explain them.

I work out an account that turns on how all of us tend to miss cues that do not tie readily to our experience in the world. There are (almost) always opportunities foregone when we take precautions, and danger accepted when we do not. Good judgment—judgment that will look reasonable when the passions of the moment have passed—has to deal with what I label the "fungibility" (between opportunities and dangers) that ordinarily confronts us. A person, and a society, needs to seek a prudent balance between the advantages of boldness and the advantages of caution. We cannot have all we want of one without giving up what is likely to turn out to be more than we want to give up of the other.

As a matter of good sense, no one really can doubt that. But cognitively, it is easy to miss one side or the other (danger or opportunity), so that fungibility is lost. Then we will be easily prompted to very firm intuitions that treat one side or the other as negligible even when that is not at all plausible as an assessment of what is actually known about the situation. This yields cases in which experts are worried but ordinary people are hard to persuade (as for some years held for seat belts, cigarette smoking, and so on). And we get converse cases where ordinary citizens are very worried about something that experts see as not very serious. I show how following up this line of analysis leads both to an account of stubborn expert/lay controversies in either direction, and also to an account of how people respond to the psychometric surveys that underlie the usual view. The argument also leads (in the concluding chapters) to suggestions about how to meliorate these conflicts, which in some ways complement but in others vary from what has usually been recommended.

2

But there is a broader agenda, which does not play much of an explicit role in the book, but which I hope will win some attention. The work reported here continues an exploration of the psychology of persuasion and belief begun in my *Patterns, Thinking, and Cognition* (1987) and continued in *Paradigms and Barriers* (1993). The focus in *Paradigms* was on famous conflicts of intuition in early science, which always were shaped by the responses of very small sets of extraordinarily qualified and committed experts (Copernicans and anti-Copernicans, Lavoisier and defenders of phlogiston, etc.). The new work reported here seeks to extend the expert/expert analysis drawn from history of science to contemporary, politically charged conflicts between expert and lay intuition.

The earlier work on cognitive puzzles and on history of science helped sharpen and deepen the "habits of mind" account of cognition that has been emerging throughout this extended project. The present book, inevitably, has a strong policy orientation. But my intent, and certainly my hope, is that the study here of a particular class of public policy conflicts will have some application to much broader questions of the character, strengths, and fallibility of human intuition: to understanding in general how we come to believe what we do.

3

The argument I make conflicts with sharp intuitions that tell most Americans not to believe anyone who claims that (for example) U.S. nuclear waste—given the sort of precautions already usual, not some qualitatively or quantitatively more demanding precautions—is not a very difficult or dangerous form of pollution. The disparity between expert opinion, which by a wide margin would support that claim, and thoughtful nonexperts, who by a wide margin would reject it, is just what makes this issue a very good one for the analysis of judgment and belief that I want to pursue.

So a striking point about the case studies in my book on Kuhnian paradigm shifts in science is worth stressing here at the very beginning. The losers in these celebrated controversies certainly exhibited what, in hindsight, looks to be extraordinary rigidity in dealing with issues that, after a while, look easy for almost everyone. Yet it was not only the losers, but also (not quite so seriously but still strikingly) winners like Copernicus and Lavoisier who were trapped by entrenched habits of mind. So it would really be naive to suppose that you and I are not vulnerable to parallel difficulties.

But the argument I am led to make puts me into the rhetorically awkward position of claiming that, on the issue at hand, my intuition (and that of a predominant consensus among experts) happens to be reasonable, and that yours (for readers who identify more readily with distrusting the experts on these issues) happens to be wrong.

Naturally, we ordinarily see our own intuitions as insightful, not illusory. Other people's intuitions may be odd, unreasonable, or whatever. But our own are what we take to be just good sense. Exceptions only arise when we also are immediately prompted to some contradictory intuition, so that we know that both cannot be right. Since we ordinarily trust our intuitions, we easily distrust—and indeed easily come to dislike—people who treat them as mistaken. But I hope it will turn out that even readers for whom the argument is not a congenial one will sometimes find it stimulating and useful anyway.

4

The study does not argue much for the appropriateness of concern about the way we regulate risk. That has been getting attention for over 20 years. A good deal of additional attention has been brought to the situation by Stephen Breyer's appointment to the Supreme Court, which in turn brought close attention to his recent book on just this theme (Breyer 1993). And then the 1994 midterm elections have enormously strengthened what was an emerging bipartisan consensus on a need for reform, parallel to the interest in economic deregulation in the 1970s, but here yielding much sharper controversy over how far reform should go.

Breyer's book is short, forceful, and carefully documented. I began my own book a few months before Breyer's was published; I completed it soon after the 1994 midterm elections. It can be taken as a psychological and sociological gloss on Breyer. There is almost no overlap in the content of the two books, so that anyone interested in the subject (even readers mainly interested in the psychological and sociological rather than in the public policy aspects of this problem) will find it worthwhile to go through his slim volume as a complement to the discussion here.

In a context where there is major emphasis on cost-benefit

analysis (as might become more common in the wake of the 1994 Congressional elections) some aspects of the reforms proposed here would become redundant. But there would remain a substantial component which, if the cognitive analysis is right, would remain relevant in such a regime. What I will call the "do no harm" proposal seeks to offset a cognitive, not a legal or administrative difficulty: a lack of a visceral sense that there *exist* relevant costs to be offset against any benefits that might come with "better safe than sorry" precautions. The substantive significance—the effect on choices—of that cognitive difficulty must be less marked in a regime that provides strong mandates favoring cost/benefit analysis (if that, in fact, emerges) compared to a regime in which such analysis is often blocked or easily overridden. But the usefulness of a "do no harm" assessment would not disappear. Within such a regime, "do no harm" would have a more modest role, but within such a regime it would also be much easier to implement.

More broadly, the general argument of the study has policy relevance far beyond the class of environmental choices, important as they are. Every knowledgeable reader will be able to think of policy areas where what we are choosing as a society runs contrary to what almost everyone with close knowledge of the topic thinks makes much sense. We spend generously on programs that virtually no well-informed person thinks can accomplish much. We are unable to muster support for other programs which promise much more, in part because such discretionary spending as is available is preempted by spending on programs that will go nowhere. And although it is never easy to get a broad consensus on what affirmatively should be done (or that nothing can usefully be done), no one doubts that we would all be better off if we could at least avoid commitments that exhaust major resources with no serious prospect of reasonable returns. So my hope, naturally, is that the close analysis here of how such dilemmas arise in the context of environmental choices will prove useful to readers concerned with parallel problems in very different policy areas.

Setting the Stage

1

At the heart of the case I want to make is a claim likely to sound too simpleminded to be believed. But after 15 years and more of controversy, what would meliorate sharp, stubborn conflicts of expert/lay intuition on environmental risk remains very much in dispute. Variations on the usual themes have been coming past the reviewing stand for a long time now. So perhaps the time is ripe for a new look.

What we want to understand is why it is so often so hard for even well-informed, sophisticated members of the general public to accept what would ordinarily be regarded as a convincing consensus of the most appropriate expert community. The difficulty has been most salient in the United States, but it is certainly not wholly peculiar to this country. The most stubborn cases have been those where the public is more worried than the experts, but it is important to notice that there are also many converse cases, in which the public is (for some substantial time at least) unresponsive to expert warnings, as for cigarette smokers and for drivers who do not bother with seat belts. And it is also useful to notice that we can point to many cases where expert/lay conflicts of intuition are well marked but little noticed. These occur where politically significant passions are not aroused even though in some such cases the lay beliefs hardly are without political significance if taken seriously.

We will want to consider some of these cases in which expert warnings fail to move the public to be as worried and hence as careful as prevailing expert judgment deems prudent. And we will want to consider why, in cases like the Kennedy assassination, public sentiment that amounts to belief that the experts are deliberately deceiving the public seems to be very widespread yet proves incapable of touching off political controversy of any consequence. Or compare the nuclear waste situation with the response to cold fusion claims or to assertions that we are being visited by extraterrestrials. The expert consensus that the latter two are not plausible enough to justify major government attention prevails quite easily despite the enormous significance of either if true, and indeed despite widespread public belief that the latter in fact might well be true.

So we can notice two sorts of converse cases: where much of the public does not believe the experts, but nevertheless there is no strong pressure for a government response reflecting those doubts, and where the public doubts the experts but the experts, not the public, are worried that too little attention is being paid to some risk. I will have something to say about converse cases. But my prime focus will be on cases where high passion is aroused about what, to experts, appear to be very modest dangers.

Even experts on nuclear waste opposed to any near-term expansion of nuclear power commonly think that high-level nuclear waste is a manageable problem, and low-level waste an easily manageable one.[1] But the public finds it very hard to accept either claim. Experts think that the Delaney Clause (which bans anything that which would increase the amount of carcinogens in foods) makes no economic sense and, in fact, probably has perverse consequences for public health. But, on the record, the public cannot believe that. We are currently spending $7 billion a year largely getting ready to clean up Superfund toxic waste sites: we spend the money (but do not actually do much cleaning up) because a cleanup that would satisfy public—in contrast to expert—

---

1. See L. J. Carter's comments on his discussions with a number of leading opponents of nuclear power, in his *Nuclear Imperatives and Public Trust* (1987), 10.

judgments of what would be reasonable would be so vastly expensive that neither the administration nor the Congress is willing to do the spending. Yet neither is able to calm the political pressure to do what so few knowledgeable people think either affordable or sensible. So we spend a few billion a year displaying determination to get on with a job few think will ever be done, and that very few experts think ought to be done on anything like the scale that public sentiment demands.

Looking across the board at such issues, they certainly take up a lot of money, attention, and effort, which, if the experts are anywhere near right, would be far more sensibly spent elsewhere. And indeed if the experts are right—and, so far as I can judge, in all these cases the expert case is the more reasonable case—we are engaged in a great deal of intense and expensive activity that will eventually come to look as peculiar if not as pernicious as the witch-hunts of earlier centuries.

The response of a good many highly regarded students of the problem is that in fact it is the experts, not the public, that is missing something. I mentioned that in the introduction, and I will consider such claims in detail in chapter 2. That the public is worried is clear. That, especially in a democracy, such worries require and deserve serious attention is also clear. But to go another step, and rationalize as sensible what seems in fact to make remarkably little sense, does not seem to me to contribute anything positive to the situation or to strengthening democracy either.

1.2

Now this study, although it will reach practical proposals in its concluding chapters, is primarily conceived as an analysis of why expert/lay controversies are so stubborn. It draws on and extends earlier work that developed an argument about how human judgment works in the context of conspicuously academic issues, such as the discovery of the Copernican account of the heavens. Here I seek to extend and apply this line of argument to the immediately political and even trendy issue of expert/lay conflicts. The earlier work dealt with general arguments about cognition and was applied in detail to laboratory experiments on judgment and then to a series of famous cases of Kuhnian paradigm shifts in the history of science. In the laboratory experiments, the issue concerns lay intuitions. In the history of science cases we are dealing with particularly famous cases of what amounts to conflicts of intuition among experts. A natural next step on this program is then the case of expert/lay conflicts. So we want to study conflicts of intuition that find lay judgments predominantly on one side, expert judgment heavily on the other, and with well-marked evidence of the mutual incomprehension that Kuhn calls "incommensurability." Expert and lay judgment see things differently, and the experts are unable to have much influence on the lay intuitions. Further, as I will try to show in chapter 2, we have no credible explanation of why that should be so.

A reader of a practical turn of mind may be tempted to skip the academic analysis and see what is offered by way of a bottom line in the concluding chapters. But the point of the main proposal—for what I will call a "do no harm" assessment as a routine component of risk decisions—is not likely to be understood well without a reasonable sense of the analysis of stubborn expert/lay controversies from which it emerges. So I hope that a reader whose main interest is in reform proposals (which I will begin discussing in chap. 7) will at least leaf through the earlier chapters. On the other hand, if you go through the argument that leads to what is, at bottom, only a modest proposal, you will find that there is a conceptual argument, backed by what I think is a good deal of evidence, to support the hope that the modest proposal might turn out to be more than just another turn of the often-played but seldom-enjoyed recommendation for improved risk communications. Since the account of expert/lay conflicts that will be developed here will be different from what you have encountered elsewhere, the specifics of the proposal for putting risks in perspective is also different. That difference that might seem superficial or inconsequential from more usual views of the difficulty, but perhaps not so if the account of how cognition works that underlies this analysis in fact captures something essential about the situation.

1.3

Since the study is tightly tied to the argument and supporting evidence of two earlier studies, I will have frequent occasion for cross-references. To avoid too much clutter in the text, these cross-references appear as much as possible in the notes. But in both the main text and in notes, *Patterns n.m* refers to chapter *n*, section *m* of my *Patterns, Thinking, and Cognition* and *Paradigms n.m* refers to chapter *n*, section *m* of my *Paradigms and Barriers*.

The central argument of *Patterns* (introduction, 1.7, 2.10, etc.) is that cognition is best understood as operating through the recognition and triggering of patterns. Human cognition is what I have primarily in mind, but the unsurprising claim is that human cognition is best understood as an extension and development of cognition in other animals. Some of these patterns, and the most firmly entrenched, are components of a species' genetic endowment. But, for human beings, the repertoire of patterns available to any individual heavily and perhaps even overwhelmingly reflects tuning to that individual's experience in the world. Even so, since much of human experience is based on encounters with the world that must be much the same in all times and places, a considerable range of the resulting repertoire (beyond the important subset that is simply innate) will be widely shared. And there will then be a further component that is shared across the community and culture in which the individual has acquired experience in the world. And there would be a still more specialized subset of patterns tied to this individual's peculiar experience, especially (for our purpose) as worked out through a person's particular role in the finely divided division of labor that marks all modern societies.

So there will be a wide range of patterns shared across the repertoires of practically everyone, a much more specialized set of patterns shared only among members of particular cultures, and a still more specialized set limited to those who are members of special segments of the culture, such as specialists or experts in certain matters. There will also, of course, be a significant range of patterns that are idiosyncratic to a particular individual. In an account of how radical innovation sometimes occurs, that individual component will play a crucial role. We want to understand how a Copernicus or Darwin or Newton comes to see things in a way that no one has seen them before, and that, at least in the short run, seem incomprehensible to most people even once they are pointed out.

But, for the work at hand, what is important is the distinction between the patterns generally shared across a community (patterns commonly firmly entrenched in the repertoires of people in the community, including members of the community who are experts with respect to some particular niche of experience) in contrast to patterns that are entrenched *only* within a specialized subcommunity of experts. Or, put another way, since our concern is with why part of a community finds it difficult to communicate with the rest, not with how some individual in a community came to see things differently from what was prevalent in the community, the individual component crucial in an account of radical discovery here plays no consequential role. On the other hand, patterns tied to the political and economic experience of people, which are only occasionally critical for an analysis of radical discovery in science, here can be expected to play a substantial role. For we are, after all, dealing mostly with controversy over public programs and obligations: that is, with issues that are intrinsically political in character.

The story we want has to be able to show how, in fact, experts are *ordinarily* able to win over the lay community to seeing things their way, but sometimes fail. So I want to account for the usual case, where lay judgment yields to expert judgment but, in a way that also lets us see why in some cases that acceptance proves stubbornly difficult, or at least stubbornly difficulty for a sufficient large part of the lay community for a sufficiently extended time to account for the stubborn and often bitter controversies that are the focus of this study.

1.4

Now the patterns that (on the view I have begun to sketch) must account for all of cognition come in two sorts. As explained in the introduction to *Patterns*, there are context patterns and there are action patterns. Suppose that, in fact, effective performance must reflect the triggering of effective action patterns—most familiarly patterns of physical action but, as spelled out in some detail in *Paradigms* (1.1–1.3), also patterns of intuition or reasoning. That could only work if we could recognize contexts where (in conjunction with other cues present or other patterns recently activated) that action pattern has proved to be effective. The action patterns governing intuitions are what I call "habits of mind," a term that by and large coincides with everyday use of such language. The stage setting or point of departure for those action patterns are what I call "context patterns," or scenarios.

The joint and interacting roles of action patterns and context patterns are best seen not as two complementary assumptions about how cognition works but rather as two aspects of the very same claim. Habits without a background repertoire of context patterns would be worthless—in fact meaningless. What could it mean to say a person has a habit but no way to recognize when the habit is to be prompted to action? What good would it be (what Darwinian story could make sense of) an ability to recognize patterns that has no consequences for what a person might do next? Even a very rich and beautifully tuned (to the world) repertoire of context patterns would be worthless without a suitable set of action patterns to move a person from recognition of a pattern to doing something ordinarily effective in the world.

But if cognition works in this pattern-governed way, then it must be the case that sometimes things go awry. A creature that never saw a pattern unless it was certainly present would forever hesitate. A creature that was never prompted to act except when the situation was exactly one it had seen before would never act at all, since circumstances in the world never exactly repeat themselves. So, in Jerome Bruner's (1956) well-known catchphrase, we are forever "going beyond the information given" when we recognize patterns and forever trying what has worked before in situations that somehow look like but are never exactly like what we have faced before. But most of the time the qualifications I have been stressing are of no practical significance. We go beyond the information given but do not get into trouble by doing that, either because what we do works or because we can notice something has gone wrong soon enough to correct our response before much has been lost.

However, counterexamples *must* occur. Sometimes a context pattern will be prompted that really does not fit the situation a person is confronting in the world, and we cannot recognize that until substantial costs have been incurred. Sometimes a habit, instead of facilitating effective action in the world, will prove to be a barrier to effective action, since the situation encountered is in some essential way different from earlier encounters with situations that look like this one. Further, situations in which such difficulties might be encountered could be expected to have a characteristic set of features—a subset of which will be the crucial concern of this study, as of its predecessors. In part we are interested for merely academic reasons in places where special sorts of difficulties appear since that is where we have the opportunity to learn the most. But also it is where these special difficulties appear that we find the clearest practical consequences of misperceptions and misapplications of a repertoire of patterns.

1.5

Consider a spectrum of judgments (see fig. 1.1) stretching from the narrow and artificial contexts of puzzles and experiments, continuing through ordinary personal experience, and reaching finally

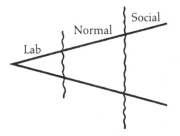

Fig. 1.1—Spectrum of experience

the broad contexts of large social issues. At the narrow end, we have limited experiences in impoverished environments. At the broad end, we have judgments that far transcend individual experience, so that the consequences of acts will often be so diffuse or delayed or ambiguous that close tuning of choices to experience cannot occur. Near the middle of that spectrum, on the scale of normal human experience, we can expect reliable tuning of acts to consequences. So we can expect cognitive anomalies to be least conspicuous in the middle of the spectrum and most conspicuous at each end of the spectrum. But nothing of consequence is at stake in situations at the narrow end. Hence, it is in the realm of social judgment, which constitutes the broad end of the spectrum, that we can expect to find large consequences linked to anomalies. But what is revealed by the artificial puzzles at the narrow end of the spectrum can help us understand what is going on in the substantively important cases at the broad end of the spectrum. For it makes no plausible sense to suppose that we have been endowed with special cognitive mechanisms for the purpose of letting us enjoy harmless illusions. Somehow, what we see in the case of cognitive illusions at the narrow end of the spectrum must operate to produce generally effective performance in the normal range, but also sometimes illusory judgment again at the broad end.

I have tried to put the habits of mind argument to use, applying it (in *Patterns* and further work included here) to several of the best-known cases at the narrow (puzzles) end of the spectrum of experience and then (in the second half of *Patterns* and in much more detail in *Paradigms*) to the general problem of theory change in science, with detailed applications to a series of particularly famous cases.[2] All of this material turns on conflicts of intuition. At the narrow end, the conflicts are between the individual's own intuition about some puzzle versus the same person's intuition about that puzzle a little later, or even essentially simultaneous intuitions about the puzzle and some logically trivial

2. *Paradigms and Barriers* treats the rise and fall of Ptolemaic astronomy, the emergence of probability, Lavoisier's overthrow of phlogiston, Boyle's defense of his work on the air pump, and the broader episode we call the Scientific Revolution.

variant of the puzzle. For the history of science cases, the conflicts of intuition are across individuals but all of whom are within a very small set of experts on the issue at hand (e.g., whether the sun moves around the earth or the reverse).

Across this wide range of material, studying what happens in terms of pattern recognition and habits of mind yields striking and coherent results. So we have reason to expect that something worthwhile, and perhaps something crucial, would come from carrying the argument into the more difficult area of conflicts of expert/lay intuition.

On the logic so far, a sufficiently clever person might have foreseen (though in fact no one was that clever) that anomalies might arise at the easy-to-study, narrow end of this spectrum that could provide clues to what goes on in the hard-to-study broad end of the spectrum. For with hindsight, we can notice an oddity that turned up in the heyday of behaviorism that suggested as much. Occasionally animals being trained to some behavior in a conditioning experiment would be moved to a new cage (the original one having been damaged in some way). When this happened, the animals would sometimes seem to forget what they had already learned. Apparently the animals learned their responses to the experimenter's stimuli in the context of the particular cage, with its particular odor, color, and so on. Cues that were logically irrelevant were being incorporated in whatever mechanism triggered the animals' responses.

Now, in Darwinian terms this odd characteristic of learning makes easy sense. Obviously, animal cognition (and, by extension, human cognition) did not evolve under the pressure of experience in artificial environments, where an experimenter was deliberately and systematically varying particular cues, and everything else was strictly irrelevant. If life worked that way, then we could expect that creatures would come to ignore cues that were part of the general background of a particular sort of experience (like the color, shape, odor of a cage) and respond only to those that were being varied, and especially to cues being varied in some salient way.

But in the natural world, since no such systematic experimenting is going on, creatures must learn many things to get on in the world, not just one thing an experimenter is interested in testing. In that natural context, creatures would profit from attending to general background cues (i.e., things changing only slowly relative to the critical stimuli in the situation) so that contexts in which a particular kind of behavior was rewarded could be distinguished from similar contexts in which the same sort of response would be irrelevant or even punished. The creature would not, and perhaps could not, discriminate between general context cues and particular reinforcing or inhibiting cues. They would all be part of the package that prompts or inhibits some response, in the way that, in recent years, has come to be intensively studied in computer simulations using connectionist models of artificial intelligence.

Since human cognition presumably descends from prehuman cognition, and if so would certainly show continuity with what we can observe about nonhuman cognition (as all other aspects of how our brains and bodies work show such continuity), this oddity of behavioral psychology suggests why even simple logical puzzles might turn out to be sometimes hard for human beings ordinarily fully capable of handling the logical demands of the puzzle. For in dealing with a naked puzzle a person is deprived of the rich context in which problems present themselves in natural settings. Of course it is only rarely that the artificial context of a set-piece puzzle prompts some striking illusion of judgment, just as it is only rarely that looking at a simple drawing prompts a striking perceptual illusion. But just parallel to the perceptual case, occasionally such illusions surely will occur. Like one of Skinner's pigeons failing to do what it knew how to do when put in a new box, a human being might miss what appear to be easy logical connections when asked to use those connections in the barren context of a naked puzzle.

Of course, *foreseeing* that illusions of judgment parallel to the long-familiar perceptual illusions might occur was certainly hard, and indeed on the record no one did see it. But once cognitive analogues of perceptual illusions were encountered in the course of experiments on human thinking and judgment, what only some brilliant insight could have foreseen was thrown in our laps. Psychologists, led by the work of Wason on logical illusions and Kahneman and Tversky's work on statistical illusions, have now been intensively studying such material for two decades and more. So a great deal of that material is now available, some of which will be exploited in chapter 3 to tease out some consequences that will play essential roles in the substantive analysis to follow.

The obvious (and frequently offered) counter to claims that illusions of judgment can show something important in actual judgment and choice is that we are only seeing peculiarities that occur when people are confronted with artificial puzzles with no consequences for their lives. Such criticism can allow that whatever is governing illusions of judgment presumably has consequences for what happens in richer environments. But in richer and more consequential environments, those effects might be very rare. Or they might only seriously affect peculiar individuals, not a society generally. Or they might be merely transient episodes, which cannot account for persisting judgments.

Yet suppose the context is difficult in some way, so that noise or novelty or absence of clear feedback or some other source of ambiguity makes reliably recognizing an appropriate context pattern difficult. Or suppose the context (even if recognized in a way that looks reasonable) cannot prompt an action pattern effective in that context because nothing that works in this particular instance of the general pattern is in the available repertoire. A person could respond only by falling back on some more general pattern (say a search pattern of some kind), which will sometimes work poorly.

Difficulties of both kinds must at least sometimes occur even in the normal range of the spectrum illustrated in fig. 1.1. But in that normal range it does indeed seem to require peculiar circumstances for such effects to be important. Yet even if illusory responses are ordinarily inconsequential in normal experience, when we are *outside* the range of normal experience such effects might at least occasionally dominate cognitive responses, and in some very significant way. In fact, even if such occasions are rare as a fraction of all experience, they will certainly be crucial on some occasion. We have to expect that at least some of those occasions will later turn out to be important.

There also happens to be a conspicuous set of contexts in which such effects might not only occur, but occur under circum-

stances that attract so much attention that they would be susceptible to detailed study and eventual confirmation. These contexts
are just those Thomas Kuhn identified in his famous study of
scientific revolutions, which introduced the notion of "paradigm
shifts" (Kuhn [1962] 1970). For revolutionary theories in science
are intrinsically beyond the range of normal experience. Yet after
discovery and exploration, these very contexts come to be within
that range, at least for scientists and engineers expert in working
with that particular class of experience.

So we can point to issues susceptible to detailed analysis but
which, at the time of discovery, had just the characteristics that
we could expect would yield counterparts of the illusions of judgment seen at the narrow end of the spectrum. And the issues are
extremely important to the people involved, who are also highly
expert in the material. In that kind of case, judgments that come
to look illusory to everyone cannot be dismissed as merely a matter of either carelessness or incompetence. Rather, the leading
figures in these episodes are always extraordinarily capable people
dealing with matters that are central to their lives. But many of
those people are making judgments that will later look starkly
wrongheaded (often even to themselves a few years later). Plainly,
the problem does not arise from failing to take the matter seriously
or from some lack of the sort of competence that you or I could
have exhibited were we in their shoes.

And indeed, when these famous scientific episodes are examined we find plenty of evidence for the sort of effects that the
cognitive view I have been outlining would lead us to expect. The
latter chapters of *Patterns* and most of *Paradigms* report on that.
In particular, I try to show strong parallels between oddities of judgment at the narrow end of the spectrum (turning on interpretation
of illusions of judgment) and oddities of judgment at the broad end
of the spectrum (turning on an interpretation of these famous Kuhnian episodes of scientific revolution). Both make coherent sense in
terms of the pattern-focused account of cognition sketched here. Together, the history of science episodes and the illusions of judgment
suggest a "habits of mind" interpretation of the expert/lay conflicts
of intuition that are the particular concern of this study. That is the
argument I want to work through in detail.

The Usual Suspects

2

The puzzle of expert/lay conflicts has been salient for two decades, and there are now many accounts on the table.[1] But in the main they fall under just three headings. By way of deck clearing, but also to provide some material that we will be able to use, consider what can be called "the usual suspects."

Theory 1 is that these expert/lay conflicts are only in form about risk. What in fact (on this view) is driving matters are deeper conflicts about power and responsibility, about human obligations to other humans and to nature, and hence about what ends public policy is going to serve. In short, the controversy is about *ideology*, not risk. Theory 2, on the other hand, allows the controversy to be about just what the experts think it is about, but the problem lies in a loss of *trust* by the public in the institutions that seek to assure it that danger is under control. Theory 3, finally, turns on the idea already emphasized, which is that what the expert sees as risk is not the same thing as what the public sees. The expert is concerned with some quantitative measure, such as expected fatalities. But the public is concerned with a far broader sense of danger that includes many dimensions beyond expected fatalities. So of course expert and lay judgment may diverge. I will call this last the *rival rationalities* view.

All three theory types come in various versions, some of

1. For a survey, see Krimsky and Golding (1992).

which appeal to people who see the lay response as reasonable and others with appeal for those who see lay responses as perverse or irrational. But the versions favorable to and skeptical of lay intuition turn out to involve only glosses on the basic arguments. Jointly, the three sorts of argument make a difficult target to attack, since they are not mutually incompatible: a person can draw on pieces of all three theories, depending on what sort of challenge is at hand. A critic consequently faces a moving target, which slips from one to another arm of the "triangular" argument. Ordinarily a person does not do that as a conscious tactic to blur direct confrontation with weaknesses. But it is very hard to avoid.

The alternative argument I will pursue turns on a claim that what we are seeing is the unconscious cuing of habits of mind that the very person involved would be likely to deem inappropriate if aware of what is governing intuition (1.5). But suppose, as will be discussed in some detail, we do not have access to such awareness. Suppose, rather that these responses are like those that account for the cuing of inappropriate patterns that yield perceptual illusions, or unconscious cuing of an inappropriate physical pattern that makes it very hard for most of us to do something so simple as tie a square knot when we want to do that. Such perceptual and physical examples are usefully (and, I have argued, properly) thought of as very close parallels to habits of mind. So, on the argument to come, it is to be expected that differences in habits of mind would turn out to be what is governing expert/lay conflicts of intuition. The general argument about habits of mind is made in chapter 1 of *Paradigms*, applied to famous cases of conflicts of intuition in science throughout the balance of that book, and put in the framework of this study in chapter 3 here.

But claims about the strong role of habits, though no one would doubt that for physical behavior, will easily seem outlandish if literally applied to thinking and especially so if applied to intuitions that have been invested with a moral force. Consequently, it may be easier to see why such an argument may nevertheless be right if we start from more familiar views. And it will help, in particular, if I can remind you not only of the usual views (the ideological, trust, and rival rationalities views already mentioned), but also put you in some doubt about how far those usual views, not only taken in isolation but also taken overall, actually explain rather than merely rationalize the difficulties we observe.

2.2

A general version of the ideological theory would go like this. Out of some combination of individual characteristics, social roles, and the characteristics of the culture and institutions in which a person has lived, a person will have some mix of orientations that (using Douglas and Wildavsky's taxonomy) can be characterized as entrepreneurial, egalitarian, and hierarchical.[2]

The *entrepreneurial* disposition is to rely on market forces, subject only to respect for minimal ground rules established by some combination of social norms and policing. The contrasting *egalitarian* tendency is to favor consensus across the group, with egalitarian rules of procedure and allocation of resources. And the *hierarchical* disposition favors control by an elite acting for the community. Then the governing value is the good of the community, not as revealed by individuals through the market or by the community as a whole through some way of reaching a binding consensus, but as seen by some elite, such as priests or nobility or Central Committee members, but also, in other contexts, senior bureaucrats, or blue ribbon committees, or the Supreme Court.

These labels are (it seems to me) best seen as indicators of a particular aspect of orientations we all share, though not with anything like equal weight either across individuals, or across con-

2. The best-known version of the ideological analysis developed from Mary Douglas's collaboration with Aaron Wildavsky, which united the skills of two exceptional academic analysts and two disciplines (anthropology and political science). But the account laid out here is just my own version, and does not pretend to be a precise encapsulation of Douglas and Wildavsky (or some variant or development, such as Schwartz and Thompson [1990] or Rayner and Cantor [1987]). The setup has roots in the grid/group analysis introduced by Mary Douglas and which led to her collaboration with Wildavsky. Characteristic social features in Douglas's taxonomy would be high group/high grid → armies; high group/low grid → communes; low group/low grid → markets; high grid/low group → street people and other segments of society that have a clear and hard-to-change status but lacking a clear sense of being a member of some community. Peters and Slovic (1995) report survey results on these lines.

texts for the same individual. In reacting to these diverse propensities we are far more constrained by our social experience than we can notice. Douglas attends closely to the ways in which these orientations (perceived by individuals as just their own sense of things) are, in fact, socially shaped.

Egalitarians care conspicuously about fairness: the clearest mark of fairness is when no one has any more access to resources or power than anyone else. The entrepreneur cares most conspicuously about efficiency and autonomy. And hierarchy is oriented to what is good for the community as a whole, but as seen from the perspective of those taken to be (or perhaps only in a position to take themselves to be) peculiarly qualified to comprehend what is worthwhile and perhaps even to embody it. So in some places and times hierarchy leads to the building of cathedrals or empires. In the contemporary world—at least in the parts of the contemporary world where readers of this study are likely to reside—it may be primarily concerned with just keeping things going.

2.3

On the egalitarian view, activities that intrinsically seem to require elites with control over enormous resources—nuclear power being the great exemplar—are intrinsically perverse. As Douglas and Wildavsky urge, the by-products of what are seen as perverse organizations and activities are pollutants, and in a moral sense, so that revulsion, not technical talk of externalities, is the appropriate response. They are pollutants not because they have particular harmful effects, although that might also be true, but because of what they are—by-products of intrinsically disgusting activities.

This scarcely does justice to the Douglas/Wildavsky argument, but it captures what I take to be the most essential point, which is to see the lay resistance to expert advice about risk as driven by considerations that lie outside a merely technocratic analysis, that is, outside the whole realm of relevance of expert judgment.

On this ideological account, if a person is disgusted by the inequalities of income and power justified by market economies, or by the alienation of individuals from the community, or by various other elements of radical criticism, then it will be easy to see corporations and large government actors (other than those whose function is to redistribute income or restrain corporations) as intrinsically evil. So of course whatever these actors want to do should be impeded: probably whatever they want to do is bad, and, even if on this occasion it is not, undercutting their claims to legitimacy and authority is a good thing anyway. And since the by-products of these disgusting institutions are profoundly polluting, it is implausible that they can really be cleaned up at all and downright wrong to suppose they can be cleaned up easily—that is, without some significant expiation of the sin they represent.

There are contrasting (left-wing and right-wing) versions of the ideological story, depending on whether such views are being presented as perverse or enlightened. Douglas and Wildavsky plainly find the egalitarian view distasteful. But they have no shortage of admirers who think they have gotten things pretty nearly right, though with the moral polarity somehow reversed. So we have both left- and right-wing versions of the egalitarian view, which need not be terribly different. Rather, the sketch just given might seem reasonable as a way to react to some people (on the left) and the same words might seem an apt characterization of a perverse way to react to others (on the right).

If I want to build a corporate headquarters or research center, I need to choose a location that will help attract the sort of people I need to staff that operation. And I will take care that the location will make the sort of impression that will appropriately influence people who do business with me. But if I am looking for a place to put a dump, I might as well put it where it is cheap to buy the land. This will not be in Beverly Hills or on the Upper East Side of Manhattan. It will be cheap where poor people live (because where else could they afford to live?) or where no one lives (where market incentives have not yet driven up the price of land). Viewed from the left, I will either be further exploiting the poor or contaminating some place where things are still in their natural state. And the right-wing response, focused on efficiency, will see this kind of argument as either naive or fraudulent, contingent on whether the adversary at hand is seen as sophisticated enough to be credibly accused of fraud on the matter.

There is beyond doubt something to a view that ideology (worldview, cosmology, and so on according to the writer) shapes what we see. If I am passionately committed to a free-market ideology, you can be quite sure that I will not exhibit intense concern over nuclear waste. And the converse if I am far to the left and convinced that redemption lies only in returning to small communities and a live-with-nature lifestyle.

I will leave it to the reader to fill in (no doubt, to taste) further details of right and left-wing versions of this ideological account. Certainly the existing literature provides plenty of help for both sides. The key point is only this: one way in which we might account for the stubbornness of so many expert/lay conflicts over risks is in the view that, at bottom, what drives the issues has nothing much to do with how an expert thinks about danger and risk. The conflict, on this view, has to do with politics, and with who is going to be given power, who is going to be denied power, and for what ends. So of course efforts to communicate expert guidance about technical details prove futile.

Taking note of this ideological account, we can readily explain why the public is always confronted with divided expert judgment. Individuals possessing technical skills but who are also strongly committed to the egalitarian view will feel most comfortable resolving all uncertainties about dangers in the way that avoids any risk that damage will be understated; their counterparts who are strongly committed to some threatened industry will feel most comfortable with resolutions that avoid overestimating damage. But if there is a general propensity to risk aversion, then those who misestimate on the side of enhancing danger will have more impact than those who misestimate the other way. And of course there are other complications. The most important is that we take it for granted that the industry analysis tends to minimize dangers, but the critical analysis is not seen as tending to maximize dangers, but only as trying to avoid underestimating them. It is not immediately intuitive, but rather takes some experience and expertise to realize how easily merely trying to make sure there is no underestimate leads to radical overestimation.

Both right- and left-wing versions of the ideological account typically exhibit some conspiratorial flavor. On the right this is needed to account for why the general public, which was happy with Ronald Reagan as president, can often behave as if committed to the egalitarian view of things; and on the left, the conspiratorial flavor is useful to explain how in fact people are ordinarily led to acquiesce in being exploited by the rich and powerful. In both left- and right-wing versions, a central role in the conspiracy is played by the brainwashing media, in one case revealing the power of left-wing reporters and in the other the power of plutocratic owners and advertisers.

All of this seems to me highly problematical as a *sufficient* explanation of the stubborn controversies we observe. On any of these accounts, it is completely unclear why the ideologically committed are so successful at rousing public opinion on some topics (such as carcinogens and radiation) but so modestly success-ful on many other matters that must be at least equally important to them. Apparently there is something about the way the public sees or responds to *these* matters that allows for a catalytic role for egalitarian partisans, and not so on many other matters. And a parallel point can be made for entrepreneurial partisans. So on its face something important is missing from the ideological story.

Further, committed proponents of either view are small as a fraction of the public, and commitments of small numbers by themselves cannot explain broad social concern. Beyond doubt, there are many people who firmly believe in UFOs, or in some conspiracy view of Kennedy's assassination. Polls reveal great public sympathy for these claims. Yet these issues have not taken hold as matters with substantial *political* significance. But other issues—questions about Americans held prisoner in Vietnam for example—have exhibited, at least for some substantial period, far more political bite. So while the existence of a core constituency that feels passionately about an issue is easily understood as neces-sary for an issue to take fire, it is not of itself sufficient. Somehow individuals not so zealously committed—people we would most

Factor	Conditions associated with increased public concern	Conditions associated with decreased public concern
1. Catastrophic potential	Fatalities and injuries grouped in time and space	Fatalities scattered and random
2. Familiarity	Unfamiliar	Familiar
3. Understanding	Mechanisms or process not understood	Mechanisms or process understood
4. Uncertainty	Risks scientifically unknown or uncertain	Risks known to science
5. Controllability (personal)	Uncontrollable	Controllable
6. Voluntariness of exposure	Involuntary	Voluntary
7. Effects on children	Children specifically at risk	Children not specifically at risk
8. Effects manifestation	Delayed effects	Immediate effects
9. Effects on future generations	Risk to future generations	No risk to future generations
10. Victim identity	Identifiable victims	Statistical victims
11. Dread	Effects dreaded	Effects not dreaded
12. Trust in institutions	Lack of trust in responsible institutions	Trust in responsible institutions
13. Media attention	Much media attention	Little media attention
14. Accident history	Major and sometimes minor accidents	No major or minor accidents
15. Equity	Inequitable distribution of risks and benefits	Equitable distribution of risks and benefits
16. Benefits	Unclear benefits	Clear benefits
17. Reversibility	Effects irreversible	Effects reversible
18. Personal stake	Individual personally at risk	Individual not personally at risk
19. Origin	Caused by human actions or failures	Caused by acts of nature or God

Fig. 2.1—Covello's list (Kasperson and Stallen 1992)

readily see as occupying a middle-ground—become polarized in a way that is comfortable for the left on some controversies, for the right on others. That is what we need to explain.

2.4

A usual complement of the ideological accounts focuses on questions of *trust*. Trust is item 12 on Vincent Covello's (1992) list of 19 factors that could account for expert/lay conflicts of intuition (fig. 2.1).[3] But placement in that list is not intended to reflect

3. See Kasperson and Stallen (1991), 133. The list was developed mainly from the work of Slovic, Fischoff, and collaborators.

relative importance. In this instance it conspicuously does not. Loss of trust is the most common explanation of expert/lay conflicts. The claim is that the public has lost trust in the huge corporate or government actors seeking to assure us that the interests of ordinary citizens are being respected.

But if one party (the public) does not believe what another is saying (here, some corporate or government agency), it is tautological that A does not trust B, at least with respect to this issue. So while it is plain that the public is not reassured by the assurances of the Department of Energy with respect to nuclear waste, it does not follow that the public disbelieves what it is told *because* it does not trust the source. An alternative explanation is that the public does not trust the source because it cannot believe the message. Or the explanation may fall somewhere in between: while distrust is real, it has major consequences on particular issues (and not so on many others) because those are the very issues about which the public would have difficulty believing assurances, even from some entity in which it had great trust.

There is plenty of evidence for at least the compromise view (where disbelief causes distrust as much as distrust causes disbelief). The most direct comes from the very data commonly cited to show that people do not trust the organizations trying to be reassuring. Polls routinely show a declining level of trust in government and corporate actors. But those same polls show a continuing high level of trust in doctors and scientists (Davis and Smith 1994, 823–48). Yet reassuring reports from distinguished panels of scientists and doctors on issues of conspicuous expert/lay conflict seem to have no great impact, which surely argues that there is something about particular issues or contexts, not simply something about corporate or government actors, that is prompting disbelief.[4] Further, if we look across issues that exhibit marked symptoms of disbelief (as we will be doing in chap. 6), public apprehension is remarkably similar even when the government

4. The American Physical Society has on various occasions organized study teams of physicists, who have the technical background to comment but no direct interest, to report on some issue of public concern: e.g., nuclear waste, Strategic Defense Initiative, and, most recently, electromagnetic fields (EMF) as a cancer risk (*New York Times*, 14 May 1995).

agencies involved are the Environmental Protection Agency (EPA) and the Food and Drug Administration (FDA), which, unlike the Department of Energy (DOE), have not had conspicuous general credibility problems.

But distrust of doctors and scientists, or of "good" government agencies, can also be defended (or interpreted, or rationalized, or excused): I could distrust what you are telling me not only because I fear you may be deliberately choosing to deceive me, but also because I fear you may be somehow under the control of institutions I distrust (even if ordinarily I trust you) or because I fear you might be sincerely mistaken. For there is never unanimity in the medical and scientific community. The public is never confronted with uniform reassurance, but only with, at most, a strong consensus. And if intense concern has taken hold across the community, then even among experts with a private view that a danger is being exaggerated, most will feel inhibited by a sense that it would be unseemly to say that too bluntly.

Yet modern technological societies are permeated through and through with reliance on expertise. We routinely assume that experts (such as the people who fly our airplanes, prescribe our medicines, drill our teeth and so endlessly on and on) know best. If something prompts us to doubt, we turn to another expert for a second opinion. If the expert opinions conflict, we look for a third, particularly reliable (we hope) expert, to advise us which of the first two experts to follow. We very rarely would consider it anything but stupid to prefer our own opinion to an expert judgment. Rather, we ordinarily have no trouble at all accepting the predominant expert view in cases where there is a clear predominance. Experts can be wrong: in fact, it is not hard at all to compile cases where experts have proved to be strikingly wrong (Cerf and Navasky 1984), and there are ample cases where a clear consensus of an expert community proved to be wrong. But what are hard to find are the kind of cases that would be relevant here: cases in which an expert consensus is in conflict with lay intuition, and it is lay intuition which in the end proved to be right.

For a long time the very best astronomers saw the world as geocentric, but that was hardly a case in which expert intuitions conflicted with lay intuition and lay intuition turned out to be better. A more plausibly relevant example was the reluctance of the French scientific community (c. 1780) to believe eyewitness reports from the provinces of rocks falling from the sky (meteors). A half century later there was the reluctance of the Viennese medical community to accept the view of women coming to their clinics that it was more dangerous to endure childbirth in the hospital wards attended by doctors than in wards attended by a midwife.

For the French case, a committee from the Academy of Sciences that included the young Lavoisier visited the provincial town after various local notables vouched for the rocks-from-heaven story. The controversy ended when they came back to report that, unlikely as it seemed, apparently rocks had fallen from the heavens. The cases we have to deal with involve no comparable strong direct experience guiding lay intuition. For the Viennese case, Semmelweiss was never able to persuade his colleagues that their failure to wash their hands thoroughly enough when moving from teaching sessions to examining patients was what was killing their patients. Wide acceptance of such views only came when Lister in England was able to make similar arguments with far greater force after the germ theory of disease had been developed by Pasteur and Koch. But there was not a prolonged controversy about the raw fact that a woman was in more danger in these teaching hospitals under the care of a doctor than she would have been if attended by a midwife.

If I had found a clear case paralleling the expert/lay conflicts that are our present concern, I would report it here. But though some reader is certain to eventually bring a case to my attention—it is hard to believe that there has never been a single good case in which lay judgment turned out to be better than expert judgment—I do not have one to report. Annals of folk medicine are probably the most promising places to look.

Ordinarily if there is an expert consensus, it dominates lay as well as professional perceptions. On issues such as whether the HIV virus causes AIDS or whether cold fusion is worth public funding, the public seems to quite readily accept the expert consensus, even though the judgement of experts falls short of unanimity and even though the very same experts are not believed on an issue like nuclear waste or the usefulness of the Delaney Clause. A report of the American Physical Society—roughly, and for several individuals literally, the same people who rather easily persuaded the public that cold fusion was an illusion—made virtually no impact with a reassuring study of disposal of high-level nuclear waste (see n. 4 above). The biomedical community, whose judgment about the role of HIV virus in AIDS wholly overwhelms the determined resistance of University of California—Berkeley professor Peter Duesberg, is still frustrated (after more than a decade of effort) at the difficulty of explaining to the public why the Delaney Clause is not a good thing.[5]

Ordinarily, expert consensus governs, and what we worry about is how the occasional consensus challenger who is right can be assured a hearing. So an explanation is needed of why consensus wins so easily in most cases, but in some minority of cases that does not happen. Then it is just a listing of the two obvious explanations (or rationalizations) of that lack of trust to say that I might mistrust expert advice on this matter for fear the expert might be lying, or for fear he might be mistaken. Neither explains why such concerns are marked in come contexts but absent in other, apparently very similar, contexts.

2.5

As we proceed, we will see a good deal of evidence for caution in supposing that distrust is causing disbelief more than the converse. But the same (or closely related) issues arise for claims about the final option: rival rationalities (theory 3). Anyone who asks people why they are worried about something that does not worry the experts quickly becomes familiar with concern over whether those experts can be trusted. But there is a by-now almost standard list

---

5. JFK conspiracy theories and belief in UFOs are cases where experts are doubted, but in a half-hearted way that does not leave politicians feeling the doubts have to be taken seriously. In contrast, consider a matter like the three-strikes-and-you're-out provision of crime bills, where a strong expert consensus not only fails to persuade the public of the folly of such a policy, but leaves almost everyone who must face an election feeling compelled to go along.

of further sources of concern, illustrated here by Covello's list (fig. 2.1).[6] And doubts parallel to those of the trust discussion arise about how far other items (beyond trust) which make up such lists actually provide reasons that *explain* conflicts of intuition, or only rationalizations that *defend* intuitions that have their roots somewhere else. This does not mean that people lie when asked why they believe what they do. That is not the issue. But an enormous body of psychological evidence tells us that a person (such as you or I) has no reliable insight into the roots of his own intuitions. Saying "I believe X because Y" is not like saying "My tooth aches," where unless you are deliberately lying you cannot be wrong. Even more remote, as will be discussed in more detail in chapter 5, is to suppose that when people are asked to rank risks on various dimensions, they will be aware of how far their responses are reflections of a marked perception of risk, as against actual causes of such perceptions.

The most transparent as well as the most pervasive illustration of ambiguity over what is cause and what is effect is provided by the trust issue just discussed. It is an automatic entry in a list like that of figure 2.1, since on any conceivable account of the matter, if people do not believe what you tell them, they do not trust what are saying. Since it is an automatic entry, its mere presence tells us nothing whatever about whether it is cause or only effect.

However, the *first* response to questions about risk is invariably that the activity or substance is dangerous. Lack of trust is then the characteristic next response to the *follow-up* question, Why worry if the experts say it is not a worrisome thing? But when the argument takes its next turn (we move to the next arm of the triangle [2.1]), discussion readily moves to the possibility that what the lay public means by dangerous is not the same as what an expert is likely to mean by dangerous. We come to the possibility of *rival rationalities*, where responses are different, but not necessarily in a way that makes one wrong whenever the other is right.

6. See Paul Slovic's chapter in Krimsky and Golding (1992).

2.6

When individuals are asked for assessments of risks on various dimensions of possible concern, intensity of worry and actuarial information on expected fatalities turn out to be essentially uncorrelated. On theory 1, it makes sense that expected harm would not be what drives these controversies, as has already been described. So it is not surprising that Wildavsky (1990), as a principal proponent of that view, was convinced that that indeed is what research on lay attitudes shows. But at least for radiation and chemicals (i.e., for just the areas where stark expert/lay conflicts are most often encountered), the leading contributor to the rival rationality view also reports that "there appears to be little relationship between the magnitude of risk assessed by experts (health physicists, epidemiologists, and toxicologists) and the magnitude of perceived risks" (Slovic 1992, 127).

Hence a basic reason for doubt that rival rationalities really explain public apprehension lies first in the absence from Covello's list of expected fatalities or any other measure of actual harm. It seems quite bizarre to suppose that expected damage is not even an important factor in lay perception of risk. It is reasonable to want an explanation of why that is so inconsequential in the psychometric work as to be missing entirely from Covello's list. Actual danger apparently rates no higher than twentieth on a list of what accounts for lay perception of risk!

But if an array of possibilities at all comparable to Covello's list is offered, then perceived risk and *anything* else (such as nearly all the items listed in fig. 2.1) are also only modestly correlated, though with a long enough list of worrying items, we can of course account for whatever we happen to observe. On the other hand, the sheer multiplicity of the dimensions that do correlate with perceived risk is also a puzzle, to be taken up in some detail in chapter 5. A study like the present one could not have been undertaken without the empirical record compiled by the psychometric studies pioneered by Slovic, Fischoff, and their colleagues. But the puzzles posed by the psychometric evidence leave open the possibility that we might do better to turn those results upside down. As with cause/effect ambiguity for the trust issue, there is no *logical* reason to take it for granted (rather than question) the presumption that lay concern is explained by the rich array of additional concerns the psychometric studies have identified. Rather, the absence of expected damage from the candidates we are reviewing raises a strong warning signal that something odd is going on. For as every reader will find from introspection and confirm by inquiring among even her most zealously committed environmentalist friends, no one in fact regards expected damage as the almost irrelevant factor these studies (on their face) find it to be.

2.7

The usual dual rationalities story elides that puzzle. It does not dwell on the absence of expected harm as a significant element of perceived risk. Rather it focusses on the extra dimensions revealed by lay responses as exposing the narrowness of expert judgment. The *puzzle*, on this reading, is not why lay judgment is deaf to what the experts are trying to say, but why the experts are blind to what bothers nonexperts.

The story goes about like this. Part of what it means to be an expert on some matter is having habits of mind that focus attention quickly and intently on some aspects of a situation and block off many other things that the expert has come to see as ordinarily unimportant or misleading. An expert tennis player learns to concentrate on the ball and leave everything else to background. And experts in general learn to concentrate on what is critical in experience with the domain at hand and ordinarily ignore anything else.

So it would not be surprising should experts have entrenched habits of mind that focus intently on expected fatalities or some related quantitative measure of risk, while lay judgment is open to a rich array of factors—such as the array (familiar in some version to all students of this controversy) displayed in figure 2.1. It then could be the narrowness of expert judgment, not the lay susceptibility to misleading (in this context) cues or heuristics, that makes experts in assessing a risk somehow blind to the significance of the richer set of concerns. So it would not make sense (on this view) to see expert lack of concern as somehow more rational than lay active concern: what we see is a matter of rival rationalities.

2.8

But we can test the rival rationalities view by looking over these extra dimensions to consider how far they might plausibly explain clear lay concern about risks that the predominant expert view sees as negligible. Do the items on Covello's list actually explain disparities between expert and lay intuition or only correlate with them? And further, do they correlate in some reasonably striking way or only as an aggregate so broad and loose that collectively they could hardly fail? In chapter 5, we will have occasion to go through the list item by item. But here a sampling will be enough.

Suppose we set aside *dread* (for a moment: I will come back to it). And set aside *distrust* (2.3), since it is automatically a reason whenever expert/lay intuitions stubbornly conflict. And trust is qualitatively different from other factors, for trust as an explanation is not inconsistent with lay judgment turning on just the same considerations as expert judgment. So trust is not intrinsically part of a rival rationalities (or of an ideological) story. But consider some other items that, like trust, might prompt us to wonder whether there may be substantial confusion between what is the cause of a perception of risk and what is a consequence of a perception of risk.

The most conspicuous candidates are press attention (item 13 on Covello's list) and accident record (item 14). These are tightly linked. For item 14 by itself hardly makes any sense. All technologies, and indeed all human activities, involve at least minor and usually also some risk of more than minor accidents. So everything would go into the "greater concern" category of figure 2.1. But perceived accidents are another matter. Suppose we tend to fix in memory incidents related to some technologies and take no special notice of comparable incidents involving other technologies. Then apparently there is some underlying propensity to see some technologies as saliently risky, not a salient riskiness that causes us to see the technology as worrisome. And since the media focuses mainly on what we seem to be interested in hearing about, that would also account for special media attention. And, in turn, that would reinforce the background propensity.

So there is great public concern about transport of nuclear waste—it would provide one of the extreme points for a ratio of expert-to-lay concerns. But as a League of Woman Voters (1985) guide allowed, the safety record for shipments of nuclear waste is "gold star." We worry a lot about nuclear waste shipments, even if the record is gold star. A nuclear waste truck that was driven into a ditch injuring nobody would be news to any editor as obviously as a similar accident involving almost any cargo other than nuclear waste would be of no interest whatever.

In 1986, a truck carrying low-level nuclear waste took a wrong turn at the Queensborough Bridge in New York and became jammed into an overpass too low for the height of the truck. There were no injuries and no leakage of radioactive material. But this commonplace accident is nevertheless familiar to anyone interested in nuclear waste transportation and was the subject of intense investigation and local news coverage. Since the event was trivial in terms of what it ordinarily takes to make news in a vast city like New York, the case is unambiguous as an example of how prior disposition to perceive danger prompted concern, which prompted intense press coverage, which (for a time) spread and reinforced concern about an incident that would have been treated as trivial had nuclear waste not been involved.[7]

Similar ambiguity of cause and effect arises on other items, even where there is a clearly reasonable basis for an influence on risk perception. Voluntariness of risks is conspicuous here, since how strongly people sense voluntariness (or its absence) turns out to be substantially subjective, and not simply a property that a computer or a human being outside the social situation could understand.

The paradigmatic example for the case of an activity with large

7. In Kasperson et al. (1987) the emphasis is on the eventual calming of concern.

expected fatalities (the conspicuous criterion by which experts rate riskiness), but where lay perceptions of risk are modest on psychometric responses, is driving a car. Everyone is familiar with auto accidents; almost everyone can recall near misses, where serious injury could easily have occurred; spectacular examples are common items of news reports. Nearly 50,000 people a year are killed in the United States. But very few people exhibit the intensity of concern about auto safety that many people exhibit about a nuclear waste shipment going somewhere in the vicinity of their city. A public that never exhibited regret at being required to wear a seat belt on an airplane for a generation resisted wearing a belt in a car. Yet even the dimmest citizen is likely to understand that the prospect that wearing a belt would do some good in the event of a crash is surely at least as good for the car as for the plane.[8]

The lack of concern about driving a car is commonly explained mainly in terms of voluntariness (item 6). Yet as a practical matter traveling by car is not really a voluntary choice for most of us. And even for the driver, many aspects of auto dangers are not voluntary in any reasonable sense at all. The obvious examples are risks from drunk drivers, who kill many people besides themselves, and highway design, which has strong and well-documented effects on the risk of fatal crashes. Both drunk drivers and less than maximally safe highways kill vastly more people than are plausibly killed by the disputed handling of radiation or chemicals. Exposing yourself to the risks of drunk drivers and less than maximally safe highways is scarcely more voluntary (item 6) or more natural (item 19) than eating foods that contain pesticide residues, but only much more likely to kill you. However, it is

8. A common response here is to say that mandatory seat belts on planes makes sense since other people might injured by someone who fails to buckle up. But considering the height of seat backs on planes, and the modest spacing between seats, how plausible could it be that a seat belt wearer would be injured by the flying body of someone not wearing a seat belt? On the other hand, a usual comment on why most people for so long resisted wearing seat belts in cars was that if the fuel tank caught fire, you could be trapped in the car. But imagine you are in a car and in a crash the fuel tank did catch fire. Would you rather be belted, so you must delay your exit by a second or so to release the seat belt, or would you rather be unbelted, but smashed against the windshield?

pesticides that we worry about and for which we doubt that government regulation can be trusted to be adequately careful.

In general, there is a large amount of arbitrariness in what is perceived as voluntary, and the same holds for many other items in Covello's list. But once we understand that characteristics like "voluntariness" in the rival rationalities argument means *perceived* voluntariness, then a very wide door is open for rival rationalities to be invoked to rationalize any response, from ignoring flagrant risks to intense concern about nonexistent risks.

2.9

Continuing on the list, familiarity (item 2) is plausible as something reassuring, while unfamiliarity is a cue to something worth worrying about. But some things seem to stay unfamiliar even after a long time. After tens of thousands of shipments of nuclear waste here and elsewhere in the world with no notable accidents, the idea of shipping nuclear wastes remains fearfully novel and unfamiliar. Understanding (item 3) makes Covello's list. But is it true that in general we understand nuclear waste storage less than we understand why the plane we are riding on does not fall out of the sky?

Other items in figure 2.1 have the character that, if indeed they are important factors, they would be more puzzling than explanatory. Why would uncertainty (item 4) be worrisome for subtle risks where the uncertainty resides in whether a nonzero risk even exists, rather than whether the risk is large? For many of the cases in which uncertainty is commonly cited (e.g., uncertainty over slight perturbations of radiation exposure relative to natural background) are of just that character. The usual account of this involves some claim that there are "many orders of magnitude" of uncertainty. But if the lower bound is zero, then any positive upper bound, however small, gives a range with an *infinite* number of orders of magnitude of uncertainty. So the "many orders of magnitude" claim may be both perfectly accurate and perfectly trivial, but still scary even when trivial.

Why would someone be more concerned about delayed risk

(item 8) than a risk that will come promptly? Would anyone actually prefer a given risk of cancer discovered tomorrow to the same risk of cancer discovered after 20 years' delay? Personal controllability is item 5 on the list. But for how many people would having a chauffeur seem daunting, since they no longer would be in control of the car?

Yet it would be a mistake to focus too narrowly on particular items, since the general difficulty with Covello's list lies not in particular items, but in something (as already suggested) characteristic of the list as an aggregate. It is truly a buffet with something to suit any taste. Are victims identifiable? Then item 10 enhances worry. On the other hand, if the victims are statistical or otherwise unidentifiable, that will also be worrisome (items 3, 4, 8, and 9 could all come into play). Is the victim me? Then items 5, 6, and 18 could be useful. But if the victim is not me, then 7, 9, and 15 will be useful. Am I worried though there is no sign or evidence of damage. Then 3, 4, 8, 9, and indeed 1 are all available. That I cannot see any effect makes sense if the danger is in the future, or by some not-understood mechanism (so I don't know where to look), or a risk of catastrophe (so although I have not seen it yet, it might be just around the corner). And so on.

Of particular interest here is the concern over fairness (obviously, item 15 but others also, notably 7, 16, and 18). For we find lots of evidence of reversals of usual ethical judgments, since ethical judgments ordinarily follow what I elsewhere call the NSNX principle ("neither selfish nor exploited"; see Margolis 1991). We ordinarily see it as selfish for an individual to insist on absolute protection of his rights and property with no regard for the costs to others or to society generally. After some point, we see a person who demands that others accept large costs to avoid trivial infringement on his rights as selfish, or stupid, even if we allow that he is within his rights. We do not see that as morally admirable, and we do not feel that it is unfair if the community is less than sympathetic to people demanding their rights when no significant harm is threatened. Eventually, social norms and even criminal law put limits on how selfish a person can be without inviting punishment. So what is fair is not always a simple matter, and cause/effect ambiguities arise also on this item. Fairness, and how ambiguities involving fairness are resolved, will eventually (chap. 5) play an important role in the argument.

2.10

A particular dilemma arises with dread (item 11). For here the problem is not just to sort out which way the causal arrow goes, but to see what the point is of including as a claimed cause something that (in this context) seems indistinguishable from the effect to be explained. Dread is easily understood as an additional and reasonable consideration in a context like that of a woman who has been raped: She might intensely want to avoid situations in which she would have to so much as think about the possibility of an attack. But even in this kind of case (and much more easily in cases of dread of snakes, spiders, lightning, flying) a person may come to see their own situation of one of excessive concern and seek to overcome it or distract it, not rationalize it.

But here, dread seems to mean only that concern about this source of risk has a visceral quality beyond logic, hence (for logically slight risks) must exceed what an expert assessment would find reasonable. But with 18 other candidates on the list, we might suppose that dread would be redundant. The other 18 items, we might expect, would explain dread. That dread itself is an element reflects the fact that even a list of 18—17 not counting the automatically available trust—does not do a good enough job of accounting for striking conflicts of intuition. We need, apparently, to add a catch-all category (dread) for cases in which everything else fails.

Suppose we try an analysis of why some marriages stay together while so many others end in divorce. And suppose that after considering 18 contributors to successful marriages so much of the variance is unaccounted for that a basket category essentially identical with just what was supposed to be explained (like "happy home") also had to be thrown in. No doubt married couples with a happy home life rarely seek divorce, and no doubt risks that inspire a high level of dread are worrisome to the public, but does either explain rather than merely describe what is under examination?

So, repeating the question with which we began this review, does a list like Covello's here, in fact, tell us something important about what drives expert/lay conflicts of intuition? Is it possible that in fact nothing in Covello's list is an important *cause* of expert/lay conflicts of intuition?

2.11

But return to the absence from Covello's list of expected fatalities, or some other usual measure of danger. Even if we take the items on the list as unproblematically causative, one might have supposed that it is when danger in its usual sense is present that lay sensitivities to other dimensions of risk would be heightened. The *absence* of the most common indicator of risk (probability of death or serious injury) might suggest that what we are looking at is not so much a list of extra dimensions that worry lay people as a list of things that might be used to rationalize lay concern in the absence of evidence of danger in its usual sense. And indeed, my own experience in asking people about why they are worried has been that items on Covello's list are brought up only in the tertiary way suggested earlier (2.4). The first response is that the stuff is dangerous, the second that the experts might be wrong in saying it is not very dangerous, and only as a tertiary response do people begin to suggest items like those on Covello's list.

So the question reasonably arises: since the primary response to riskiness is to mention danger (meaning, if you ask, what you would expect, i.e., danger of killing people, of causing cancers, etc.), then why is that usual focus of expert analysis nowhere on Covello's list at all? Douglas and Wildavsky have argued that cultures select risks to serve other purposes than avoiding physical harm, and there is, no doubt, some merit in that claim. All the same, it seems to be going too far to suppose that physical danger has nothing to do with what is perceived as dangerous.

But a possible key to this puzzle comes to hand if we attend to the possibility that "risk" and "danger" (or their adjectival forms, "risky," "dangerous") are used, as indeed we use language generally, in context-dependent ways. In formal logic, language such as "or" or "if . . . then" is given an unambiguous meaning that is rigidly adhered to. That is essential for various technical purposes, but it is not the way human beings (including human beings exquisitely well-trained in formal logic) ordinarily use language. Rather, we all very fluently but ordinarily wholly unconsciously rely on contextual cues to see what, within the range of meaning and nuance a word might convey, is the sense that makes sense right now. It is only in odd cases that fluency in apprehending a particular meaning comes to explicit attention (as illustrated in *Patterns* 5.4). Since words in general smoothly take on a contour that fits the context at hand, naturally we should expect that words of special concern here, such as "risk" and "danger," also work that way.

Now one thing that risk often means in ordinary language is something like "more risky than I feel comfortable with," where what prompts that sense of discomfort and prompts enhanced vigilance is something exceptional about the context. For since we live in a world that is full of risk, if we had that feeling of discomfort whenever there was risk, we would never feel anything but uncomfortable. Obviously that is not, and indeed could not be, how things work. Rather, as is generally true we make our adjustment to usual situations, and the feeling of discomfort is reserved (not consciously, of course, but ordinarily productively) for situations that in some way are unusual (*Patterns* 6.4).

But danger or risk in a statistical sense could not be irrelevant to what prompts that state of enhanced vigilance: rather we would expect that, ordinarily, how easily we are prompted to a state of enhanced vigilance and how long the state of vigilance is likely to be sustained will be clearly influenced by how really dangerous, in a statistical sense, the situation is. But the statistical sense of risk now becomes a conditioning factor, not any longer a definition, of what governs the *visceral* "vigilance" sense of risk.

We would not get along well if we could only respond to actual danger, since very often we cannot directly see oncoming danger until it is too late. Rather we must respond to cues that somehow have come (ordinarily subconsciously) to signal danger. And the perceived sense of danger will be relative to what we are habituated to. A person who has gotten into the habit of wearing a seat belt while driving is likely to have a well-marked visceral sense of danger when driving without a seat belt, while a person without that habit has no visceral sense of risk whatever in the same situation.

More needs to be said about this cuing process, which in fact is the central issue for the account I am developing. That discussion will begin in chapter 3. But it is worth stressing immediately that cues that prompt us to vigilance—that activate a visceral sense of risk—must be those that are effective in ordinary experience. The alerting process will have the potential for prompting odd responses in contexts that are ambiguous, or noisy, or far from the common experience in which those cues had proved reliable in tuning responses to contexts. For in all those cases it is inevitable that ordinarily reliable alerting signals will sometimes be only misperceptions or oddities of circumstance (1.5).

In any case, we now have explicitly on the table *statistical* "danger" or "risk" in a detached sense in which it is something that we could ask another person to investigate and report back to us what it is (say as a particular probability of injury), and we also have before us *visceral* "danger" or "risk" in the discomfort/ vigilance (affective) sense where it would be chancy to ask someone else to figure out what it is, since what is involved is a subjective sense of uneasiness about whether the risk (in the objective sense) is being adequately attended to. I could ask someone else to tell me whether a given risk is strong or weak in that subjective, visceral sense only in the way that I could ask some else to tell me whether a joke is funny. Their report will be their visceral response to the punch line, which might or might not bear some resemblance to mine. I might find a situation hilarious, though Queen Victoria was not amused.

For our discussion, it is going to be essential that we keep these two senses distinct: the visceral, subjective risk (the affective "vigilance" response), and the objective statistical sense. Note well that so far I have not used the labels "lay" and "expert" in this discussion. *For it would be misleading to do so, since what distinguishes visceral risk from statistical risk is not at all an expert/ lay distinction.* For an expert just as much as for someone who is not an expert, a judgment about whether we are being cautious enough about some risk is ordinarily a visceral sense of what response feels right.[9] Since visceral risk reflects enhancement of our routine state of vigilance, obviously visceral risk cannot be reduced to statistical risk, even for an expert in statistical risk—or, more accurately, for an expert in the particular statistical risk at issue since no one is expert in all statistical risks.

Now it *might* be—that is, logically it could be—that what makes an expert relaxed and another person vigilant about a particular statistical risk is just what rival rationality supposes: the lay person is concerned with many dimensions of risk, while the expert focuses intensely just on expected fatalities (or something like that). Hence with respect to the same statistical risk, the expert's visceral risk may be radically different from the visceral risk of an ordinary person.

But although distinguishing between statistical risk and visceral risk does not refute rival rationality, it does make it easier to see that rival rationality *might* be illusory. For visceral risk is, by definition, an intuitive sense of things, like seeing a joke as funny. It reflects a sense of more than normal apprehension or vigilance or wariness in connection with some statistical risk. Something about that statistical risk must be cuing that enhanced visceral risk, but a person has no reliable introspective access to what it might be. However, given enhanced visceral risk, we would presumably see that reflected in an elevated response to the items on Covello's list (and the converse for oddly low perceptions of visceral risk). Then, as I have been urging on other grounds, it would be an open question, and not at all something to be merely taken for granted, whether the candidate items are better described as characteristics that enhance vigilance or as characteristics that might be enhanced *by* a sense of vigilance. I will take that up again, and in more detail, in chapter 5.

Certainly we could expect the most persuasive resolution of this ambiguity of cause and effect for Covello's list will vary across items, and in a way that itself would vary across different sorts

9. I am tempted to write "can only be" rather than "ordinarily" here, but we need to allow for cases in which some standard is in effect and a person is just looking at the numbers to decide whether more caution is required. In that case it is not a matter of how the situation feels.

of target entities or activities. "Voluntariness," to mention the most obvious case, and as will be discussed in more detail in chapter 5, is certainly sometimes an unambiguously causal element in a person's sense of visceral risk. But in various contexts, as I have suggested, it seems that *perceived* voluntariness is driven by a sense of visceral risk, not in fact the cause of it.

Taking note of that possibility of cause/effect ambiguities for these added dimensions, we should be less surprised that psychometric responses do not correlate closely with statistical risk (2.6). For expert/lay controversies are mostly about statistical risks very much smaller than risks we ordinarily notice or indeed have any way to notice. As I will try to show in chapter 5, it is under just such conditions that it would be hardest for cues necessarily acquired mostly from everyday experience to function in their ordinarily reliable way. Hence it is also for just such risks that it would most easily happen that perceived risk—the visceral sense of risk that is really at issue here—could show little connection with statistical risk, as if "kills few" versus "kills a lot" or something of that nature were not important for lay perception of risk.

2.12

Now review the overall argument of this chapter. We have considered the rival rationality claims of theory 3, and also the trust theory 2 and the ideological theory 1. These usual suspects have *something* to do with what we see. But if we try to pin down how far they could account for the puzzles that concern us, not only does no one theory appear nearly strong enough to carry that burden, but neither does their combination. The ideological account is important because the catalytic role of ideologically committed activists seems to be an essential piece of the story. But it does not at all account for strong feelings far more widely shared than the ideology that might account for them. Trust is an automatic entry, given disbelief of what the experts say. And rival rationality turns out to be such an all-purpose—even every imaginable purpose—list of highly malleable possible contributors to concern that we can very sensibly doubt that it really explains much, though it can rationalize anything.

It would be unrealistic, even unreasonable, to suppose that this chapter will *persuade* a reader (not already partial to the view) that the usual suspects do not in fact adequately account for expert/lay difficulties. But I hope I have opened the door to the possibility that something is needed beyond even the combined ideological, rival rationality, and trust accounts. Perhaps you will be prepared to consider a different account, even an account odd enough to overlook until after the more reasonable, and by now the usual, suspects had been thoroughly tried.

How Habits of Mind
Govern Intuition

3

The intricate division of labor that characterizes modern life guarantees that much of what is going on at any moment is hard to grasp in any deep or reliable way except by people with specialized experience. So it is in no way surprising that my casual lay sense of what feels dangerous or safe in some context—say a medical or automotive or electrical context—will sometimes be very different from that of my doctor or auto mechanic or electrician. But ordinarily, once the appropriate expert tells me I have a wrong impression I listen to the expert. The cases we need to explain are just those where the expert tells me I have a wrong impression, but somehow that does not really impress me. My reaction instead is to feel doubt about whether I should trust that expert.

So we would like to pin down what might distinguish what are certainly commonplace responses, in which we follow an expert consensus if it exists (and indeed consider it stupid not to do so), as against the cases, on the whole less common but politically and socially conspicuous, in which we find it easier to suspect that the experts might be mistaken or corrupt or otherwise untrustworthy. The point, of course, is not that experts are never wrong or corrupt or otherwise untrustworthy, but that we ordinarily do not ignore expert advice. Rather, in the absence of some special incentive (the decision is very important to us, or we fear a conflict of interest), we do not actively worry about it. So we might ask for a second opinion if the auto mechanic recommends an expensive repair. But if a second mechanic—one who will not get to do the work—also says that job is needed, we are unlikely to hesitate further.

I will spell out the main argument I want to make about these exceptional cases in chapters 4 and 5. But here I want to put on the table several more general claims about how intuition works.

A warning is needed before we start—an important enough warning that I will repeat it with some further detail at the conclusion of the chapter. It is convenient to study how intuition works by looking closely at cases where it works badly and then contrasting such cases with related cases in which intuition works well. The cases that tend to seize attention, but also the cases from which we can learn the most, are ones that exhibit really striking illusions. And indeed, a major result of such study is an enhanced awareness that the processes that yield our intuitions can produce illusions as well as insight. But the claim is not at all that those processes *ordinarily* produce illusion rather than insight, or that an ideally rational person would try to ignore intuition and believe only what she can derive from formal reasoning. I have developed that point at length (e.g., *Patterns*, chap. 4; *Paradigms*, chap. 10). Yet illusions do occur, and in some contexts illusory intuitions may be a central feature of what is going on.

I will develop the argument in terms of the "scenario" account of illusions of judgment developed in *Patterns*. Inevitably, I have particular affection for my own account. But that argument is only one of many accounts contending for attention. The overall argument of the study is not contingent on one particular account of illusions as against various others. All that is really essential is the well-established fact that various simple puzzles routinely prompt even highly intelligent, highly sophisticated individuals to sharp, confident intuitions that turn out to be wrong.[1]

3.2

I start by laying out four key points, beginning with a claim that is not likely to seem controversial though its force is hard to

1. Many surveys of this work on cognition and intuition are now available, e.g., Hogarth (1987).

appreciate without looking at concrete illustrations. I will give an exemplar shortly. More detailed discussion and further illustrations are provided in *Patterns* and *Paradigms*.

Point 1.—We ordinarily have a powerful propensity to see a clear intuition as right. We do not, however, have much occasion to notice that. For we can notice that propensity to overconfidence in our own intuition only when things go demonstrably wrong. But it does not often happen that we have a sharp intuition that is certifiably wrong—so clearly wrong that we eventually would come to see it as wrong ourselves—*and* someone or something in the context draws attention to the intuition and to why it is wrong.

Suppose, as is most often the case, we have a clear intuition that is not challenged. Then we do not think about our confidence at all. Or if an intuition is challenged, but not so effectively that even the person who has the intuition comes to see it as wrong, then we just see the conviction we are right as a confirmation that in fact we are right. We are all having intuitions continually— FOR example, as you read this, that the capitalized word in this sentence is "for." But you do not notice any confidence in that intuition, since if no glimmer of doubt is prompted, no such assessment is prompted. Indeed a creature whose cognitive apparatus worked otherwise would be crippled, overwhelmed by continual interruptions of thinking by thoughts of "Oh dear! There's another correct intuition." It would be only if I challenged you and claimed you were suffering an illusion—the capitalized word is really "cat"—that you would be prompted to a sense that I cannot be right: your intuition that it is "for" is so clear that you are sure it could not be wrong.

But you would not think it demonstrates anything that you have the conviction you could not be wrong in a case where in fact you are not wrong. So the claim of this point (that it is hard to make us doubt our own intuitions), although perhaps not controversial, needs illustration with some actual intuition that comes to be seen as uncontrovertible wrong.

And a corollary to this first point has already been mentioned in passing, but it warrants distinct emphasis:

Point 2.—It is only when an intuition is challenged by some friend or adversary or observer (as I did in an artifactual way with

FOR a moment ago), or by something unexpected or anomalous in the situation, that we are prompted to examine our "second-order intuition" about the possibility that our primary intuition might be wrong. *Overcoming* a clear intuition requires some new feature in the context which provokes a conflicting intuition of at least equal clarity: it requires a *rival*, in the sense of *Patterns* 7.6. In the context I will use to illustrate these points, providing such a rival is not hard in any absolute sense, but it turns out to be remarkably hard relative to the complete triviality of the issue to be resolved.

Point 3.—In particular, you are likely to notice that merely providing a logically clear counterargument need not be sufficient to dispel the mistaken intuition—even for a reader as intelligent and sophisticated as you are. Hence, if the counterargument is not absolutely clear (if the argument were open to some dispute, even if not much dispute, or the argument requires expert experience to fully appreciate) then overcoming a clear intuition merely by cogent argument may be exceedingly difficult. The environmental controversies that concern us are conspicuously subject to those adverse conditions. So it is essential to attend to both aspects of the situation: (*a*) it takes a cognitively effective rival intuition to challenge an intuition, *and* (*b*) mere logic (by itself) can be remarkably insufficient in filling that role.

Point 4.—Finally, we have no introspective access to what prompts our intuitions, so that some sense of things may be guiding my intuition even when I would deny that could make sense. My confident intuition about what is reasonable could reflect tacit inferences that I would regard as wholly *un*reasonable. Since the force with which habits of mind covertly govern intuition is hard to appreciate, it is easy but a severe mistake to suppose that this could only happen to stupid or naive people, not to people like you and I. But as I have already urged (1.6), in contexts that are unfamiliar, or impoverished, or difficult in some other way, the very habits of mind that account for our fluent and effective performance in more normal circumstances can yield blunders of intuition that look confidently right to an otherwise highly intelligent and sophisticated person. And that we cannot see what is prompting an intuition means that diagnosing and repairing an illusory intuition may be very hard. But this and the previous points need concrete illustration.

3.3

Here is a puzzle that will usually prompt a response that illustrates each of the points just made. And if you happen to be in the minority for whom this puzzle fails to stimulate a cognitive illusion, your disappointment could almost certainly be relieved by some other examples. Here is the general setup for the "chips" puzzle:

> Three poker chips are in a cup. One is marked with a BLUE dot on each side, and another with a RED dot on each side. The third chip has a BLUE dot on one side and a RED dot on the other. So there is one blue/blue chip, one red/red chip, and one blue/red chip.
> Without looking, you take out one chip, and lay it on the table.

I will be using that setup in several different ways to illustrate various points, starting with this set of questions:

1. Suppose the up side turns out to be BLUE? What is the chance that the down side will also be BLUE?
2. What if the up side is RED? What is the chance that the down side will also be RED?
3. Before you see how the chip has fallen, what is the chance that it has the same color dot on both sides?
4. Suppose you answered 1/2 in response to questions 1 and 2. That would mean that whichever the up color of the chip, the chance is 50/50 that the color on the down side is the same. But if, at question 3, you said that chance is 2/3, are you contradicting yourself?

The usual response to questions 1 and 2 (Q1 and Q2) indeed is 1/2. As implied by Q4, the usual response to Q3 is 2/3. And the response to Q4 from the large majority who report 1/2 for Q1 and Q2 but 2/3 for Q3 is almost always that there must be some mistake in the reasoning that claims to show a contradiction.

Indeed this is very often a most emphatic response, which is hard to overcome: all the more so for people whose experience in the world gives them confidence that they could not have a *mistaken* clear intuition about such a simple matter. Nevertheless, you will eventually come to agree, however strong your intuition otherwise is right now, that the correct answer is always 2/3: the powerful intuition that 1/2 is right for Q1 and Q2 is an illusion.[2]

But suppose that, at the moment, you still do not believe that. Then if indeed you are eventually persuaded that the contradiction is there, we would certainly have an illustration (on a small, but also on a convenient-for-analysis scale) of the four points developed in the previous section.

Reviewing: (1) Until an intuition is challenged, it does not even occur to us that it might be wrong. (2) Even if an intuition is challenged by a rival intuition, the conflict must be very stark and the rival intuition must be as solidly entrenched as the one under challenge. Here the 2/3 intuition about a sensible response to Q3 in fact contradicts the 1/2 intuition for Q1 and Q2, but when confronted with Q4, the contradiction, which will surely come to seem stark eventually, is (for a very large majority of people) not yet stark enough to be effective. (3) Mere logic alone will rarely be effective in challenging intuition. The logic of the counterargument given in Q4 is very simple. So if it is right, which in fact it is, then it provides a good illustration of the point that a clear logical counterargument is usually not enough to overcome a clear intuition: our response tends very strongly to be suspect instead that something must be wrong with the argument. And (4), if you are still stuck with a clear intuition that the right answer to Q1 and Q2 is 1/2, then obviously you cannot be conscious of what is prompting an intuition you will soon be agreeing makes no sense.

2. Think of it this way. Suppose you close your eyes and draw a chip and it is now on the table covered by your hand. From Q1, if the up side is blue, the chance is 1/2 the down side is also blue. And from Q2, the same holds also if the up side is not blue, but red. So even before you know what color is on top, you would know the probability the top color will be matched by the bottom color is 1/2. But from Q3 (if you gave the usual answers) you would simultaneously believe the probability is not 1/2 after all, but 2/3. So (Q4) are you not contradicting yourself?

3.4

Given a strong conflicting intuition about the chips puzzle, the modest—in fact, quite trivial—logical steps needed to go from Q1 and Q2 to seeing the contradiction with Q3 are enough to lose even highly intelligent subjects. What is needed, as mentioned, is a *rival* intuition that can jar a person into *seeing* that he is believing contradictory things. In a real problem, such as the environmental conflicts we want to understand, that can be terrifically difficult: for it might require concentrated effort to master enough of the complexities of the situation to be able to see a contradiction. A person may not have the resources needed to do that (at a minimum it takes time); an even larger problem is finding the motivation to work at seeing how something you confidently believe might be wrong.

For our chips example, providing an effective rival is not so difficult, since the situation is so trivially simple. We can push the two questions so tightly together than only a one-step argument is needed to lay bare the contradiction. And we can do that in a number of different ways. And we always have available, as a final resort, challenging a person to put three chips in a cup and to just see for themselves what happens.

Here are some alternative ways of framing the questions: logically these are *exactly* equivalent to the first set. But cognitively, seeing the contradiction sometimes becomes easier. Coming next, Q1&2 is just the original Q1 and Q2 (which everyone answers the same) combined into one question.

1&2. What is the probability that the color on the bottom will match the color you see on the top?

If your answer to Q1&2 is 2/3, you may now be able to see the contradiction between a 1/2 response to Q1 and Q2 and a 2/3 response to Q3. Q1&2 is logically identical to Q1 and Q2 put together. (It says, "if the color you see on top is blue, what is the probability the bottom is blue?" *and* "if the color on top is red . . . ," etc.). Unless you gave different answers to Q1 and Q2, which in fact no one does, then the answer to Q1&2 logically should be the same answer as for Q1 and Q2 posed separately.

On the other hand, Q1&2 is also logically identical to the original Q3. But cognitively we can see it is closer to Q3, because people ordinarily give the (correct) 2/3 response to Q1&2 though almost everyone (incorrectly, but with great subjective confidence) answers 1/2 to Q1 and Q2.

Alternatively, sometimes the next question will provoke a sufficiently direct awareness that the 1/2 answers must be wrong:

4′. Suppose you do the experiment 100 times, keeping track of how often it turns out that the color you see on the up side of the chip will be matched by the color on the down side? How many times (out of the 100 tries) would you guess that would happen?

And a pair of questions that may make the contradiction even easier to see are these:

1*. Suppose the up side turns out to be blue? What is the chance that the down side will also be blue?
2*. Among them, the three chips have three sides with blue dots. Is there a way to put numbers on the blue dots in such a way that number 1 is twice as likely to turn up as either number 2 or number 3? Yes or no?

If at Q1* you correctly think the blue/blue chip indeed occurs only half the time when the up color is blue, then obviously you could just number the blue dot on the blue/red chip "1" and the two dots on the blue/blue chip "2" and "3." Then you would expect to see "1" as often as "2" and "3" combined. So the answer to Q2* would be yes. But confronted with that absurd conclusion, you may see that something must be wrong with a 1/2 answer to Q1*. Yet even then most of us intuitively continue to *feel* right about the 1/2 answer, even though by now (perhaps) we know it is wrong.

3.5

A considerable menagerie of problems like "chips" have been noticed in recent decades; these cognitive illusions are entirely comparable to the collections of perceptual illusions and gestalt reversing drawings discovered by earlier psychologists. Just the point of introducing these puzzles here is that what can be learned by close attention to how we respond (I want to argue) can give important insight into what happens on large issues of social controversy.

If you indeed have been prompted into the illusory response, you have fallen into the contradiction even though you had been elaborately warned at the start that you were about to be exposed to a cognitive illusion. And some readers are still finding it hard to believe that the correct answer to Q1 is 2/3, even though I have been, as boldly as possible, assuring you that eventually you will be ready to admit that you are contradicting yourself.

On the other hand, it is important to note that my confidence in the reader's eventual discovery of self-contradiction relies on the fact that in this case—but ordinarily not so in real-world counterparts of this situation—the logic involved is trivial *and* there is a simple way to satisfy yourself about what is really the answer that works: a way that requires no logic at all. The simple way is just to do the experiment. Then you can count results and see for yourself whether the chance that the color on the bottom matches the color on top depends on whether you look at the color on top before you check the color on the bottom. Indeed, by now perhaps a thought experiment will be all that is needed. For how could you avoid seeing a color on top before you know what is on the bottom? If that is all it takes to make the blue/red chip come up half the time, then what other outcome could you ever encounter to a series of such draws?

3.6

Showing a person why the usual illusory response to "chips" is indeed illusory is usually slow going until (as in the second and third versions I have given) the questions are pushed so close together that barely more than a sentence is required to show the contradiction. So it is not just that virtually no one caught by the illusion spontaneously sees they have taken a logically impossible position: beyond that, unless things have been carefully arranged, it remains hard to see the contradictions even after it has been

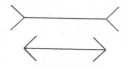

Fig. 3.1—Muller-Lyer illusion

pointed to. Nor is this because only a person trained in probability could follow the argument. On the contrary, the people most stubbornly trapped by this illusion tend to be people confident that they could not possibly be wrong about a probability problem this simple.

A key point for understanding the stubbornness of the illusion (and point 4 in sec. 3.2) is that in general we have no access merely by introspection to the cues that prompt an intuition, so of course we have no access to what prompts an intuition that is illusory. This is not at all a merely supplemental point. If we could *see* what is cuing our intuition about what makes sense here, the chips illusion still might occur, but it could hardly persist. If I could see what tacit sense of context prompts my intuitions, you could directly challenge the appropriateness of my responding to a context that plainly is not appropriate. And in a case where the situation is clear, I might agree with that, so that (as with a perceptual illusion) even if I still have the cognitive illusion, I know it is wrong.

But since I cannot see what is prompting my intuition, if I have not yet in some other way come to see my intuition as wrong, then I might actively agree with your claim that a certain context is inappropriate, but nevertheless continue to be confident my intuition is right anyway. I do not see my intuition as covertly contingent on what I am explicitly ready to deny.

With a perceptual illusion, the most familiar example being the Muller-Lyer illusion (fig. 3.1), it is much easier to see how we might isolate the controlling cues; and then it may become straightforward to conjecture the misperceived context they prompt. Our eyes tell us the upper line in figure 3.1 is longer, but once that perception has been challenged, we all know how to use a ruler to check what is really there. And we can see that the illusion disappears as we make the arrows, hence the (false, in this context) perspective cues provided by the arrows less emphatic. Alternatively, we can make the width cues more effective by pushing the two lines closer together. The angles seen are like those that ordinarily correctly cue us to see one object as more distant than another, and our visual machinery is such as to make such an object seem bigger than the physical size on the retina alone would imply.[3]

In a natural situation, though, it might not be possible to check the perception, and in particular it might not be possible to check it in time to preempt a dangerously mistaken response. Surely many accidents result from illusory perceptions, but since accidents happen fast and we do not get a chance to rerun them so the victim can take better note of just what he thinks he is seeing as the event occurs, we do not easily get information to adjust or correct our impressions.[4] Imagine the situation if my first encounter with the Muller-Lyer setup was as a pair of lines on a distant wall, seen through field glasses. And suppose you (the expert) had the task of persuading me—just by giving me verbal arguments—that what I see so clearly is only an illusion.

If we somehow come to be ready to believe we are suffering an illusion, as for everyone for figure 3.1, and (by now) for nearly all readers for the chips puzzle, then our second-order intuition (intuitions about our intuitions) is that something we cannot seem to consciously control is pushing us into seeing something we now suspect is wrong. So we are open to suggestions about what that might be. If you are now convinced that your initial response to the chips puzzle was illusory, you will be open to and indeed curious to hear a suggestion of what accounts for that, as you would be open to an explanation of why you see the Muller-Lyer drawing in an illusory way. But if you were not already persuaded that your intuition might well be mistaken, what claims to be an explanation of why you are mistaken will easily be, on the whole, somewhat less than useless. More likely, it would just make you angry.

3. For a detailed discussion, see, e.g., Frisby (1979).
4. One case, though, where things are well-documented is that of pilot error (see Kahneman and Tversky 1984).

3.7

Whether a habit is perceptual, or physical, or (as is our concern here) a habit of mind governing intuition, we not only have no access to what is cuing a response, we ordinarily have no awareness that there is any such process. We just see what we see, or do what we routinely do. Until it is pointed out, virtually no one is aware of why they so often tie a granny knot when a square knot is intended.

Difficulty in tying a square knot, which I once thought a personal quirk, is commonplace (2.1). On the habits of mind view, that should be due to some very solidly entrenched physical habit for tying granny knots. Tying shoelaces then is the obvious candidate for the source of this very stubbornly entrenched disposition. And indeed if you carefully attend to how you tie your shoelaces, you will find it is as a granny knot.

But almost no one is aware that shoelaces are tied as a granny knot. We all do it without knowing we do it, which provides a trivial but effective illustration of how habits become tuned to experience. For if you tie your laces as a square knot, then the lower knot will be tightened when you pull apart the loops; but if you tie them as a granny, the lower knot loosens when you pull the loops. So we all (tacitly) learn to tie our laces the way that works well, and we are inhibited from tying the alternative knot that gets us into trouble. But since all this is entirely unconscious we are then puzzled that, when we want to tie a square knot, we so often get this simple task wrong.

This difficulty with a trivial task (tying a square knot) then provides a simple demonstration of the power of a strong, and wholly unconscious, habit to control our response even in a context where (because we have often erred in the past by tying a granny when a square knot was wanted) we know we are vulnerable to error. Naturally (I argue) in a context in which it is harder to recognize what response is appropriate, it will be harder to avoid the dominant habit and harder to recognize that a correction may be needed. On the other hand, with a theory in hand that tells us what to look for, it may not be terribly hard to uncover a plausible conjecture about critical cues to some familiar but here mistaken context *or* what will play a still larger role here: how the absence of strong cues normally present when some response is appropriate lets a default response take hold.

The habits of mind argument tells us to look for patterns of entrenched experience that could come to be treated as mutually exclusive alternatives, such as the "fight or flee" response of animals. For the square knot case, this tells us to look for some common experience that looks like but is somehow an incompatible alternative to tying a square knot. The obvious candidate then is tying shoe laces. And a simple experiment with going through the motions of tying your shoelaces (but this time leaving out the loops) confirms that indeed tying shoelaces is tying a granny knot.

The habit that prompts the chips illusion seems to be this. Sometimes things happen in the world in a way such that what happened last tells me something about what will happen next (*the correlated case*). And sometimes what happens next is random with respect to what happened last (*the independent case*). The correlated case itself provides a further pair of rivals, depending on whether the correlation is positive or negative. The former is the case where induction works (supposing the future will be like the past); the latter is the case where I would do better to suppose that things will be different next time. Life offers many encounters with both kinds of situations, so that a creature that learns to discriminate in a generally reliable way between them will do better than a creature that fails to do that. But the "default" case is likely to be the neutral (independent) case, which is consistent with the propensity prompted by the chips puzzle.

The chips problem seems to provide a case (its close kin is the better-known Monty Hall puzzle) where we are somehow cued to see the situation as independent, though the language plainly describes a correlated situation. If the signal provided is that the dot on top is blue, then (since this signal is on both sides of the blue/blue chip, but only one side of the blue/red chip) that affects the chance that you have gotten the blue/blue chip. Seeing the blue/blue and blue/red as still equally likely (as they were before you got that information) means you have missed that correlation. This could be because there are normally reliable cues present that happen to be misleading on this occasion. But it also could be (as mentioned already) merely that the limited array of cues that characterize a naked puzzle (in contrast to natural situations) allows a default to take hold. For the default's virtue (what gets it into the position of being the default) is that, if you are unclear about what to do, that is probably the most prudent thing to try. But if the context is not what you ordinarily encounter, then even if the information that logically tells you the default is wrong is in front of you, you may miss its significance and choose the default anyway.

For the chips problem, the probability you will draw a same color chip was 2/3, and the probability you draw the single, mixed color (blue/red) chip is 1/3. This gives a correlated, not an independent, probability when you see what color is on top. You are not choosing randomly between the blue/blue and blue/red chips so that each is equally likely; instead you are guessing the color on a chip that was already chosen, where one of the possibilities has a blue dot on both sides but the other has a blue dot on only one side. That it is hard to see the significance of this distinction—nearly everyone at first misses it—suggests that we are seeing the effect of a well-entrenched habit of mind (here making the independent case the default).

And indeed, suppose that our experience in the world is such that we more easily get into trouble treating events as correlated (when they are not) than with treating events as independent (when they are not). Then in an unfamiliar context (such as that of an artificial puzzle but also perhaps in vastly more consequential ones) it may not take much to cue us into an inappropriate context. In fact, even with information on the table that logically should override the default, the mere absence of well-established background cues ordinarily present when a nondefault response is appropriate, may lead the observer to act as if the case is independent when it is actually correlated.[5]

---

5. For most of us, physical situations are rare in which tying a granny knot gets us into trouble, but tying shoelaces with a square knot will always get you in trouble when the time comes to untie the laces. If you were a sailor, however, it would not take long before you recognize contexts in which you routinely use a square knot, and in those now well-entrenched contexts, you would soon be tying the square knot reliably without giving it any special thought.

3.8

What is important for the balance of this study is not the details of the chips puzzle, but the more general points that this puzzle can illustrate. Reframing the points of 3.2: (1) Absent a challenge (from anomalies in the situation, or from an adversary) it does not even occur to us that an intuition could be mistaken; (2) even if there is a challenge, until we are focused on some anomaly whose significance itself is well entrenched in our own experience, our response to a challenge is to see the challenge, not our own intuition, as wrong; (3) logic alone is remarkably ineffective unless the target of that logic is motivated to work at attending to its significance; and (4), we have no direct insight into what cues are guiding intuition nor indeed normally any awareness that there is any *process* involved. Subjectively, it feels like we are just directly perceiving the way things are. So arguing that it would be inappropriate to respond as if the situation were X does not change any minds even when the person agrees with that, since the person does not see himself as responding that way.[6]

3.9

A reader might now want to argue that the particular explanation proposed (turning on rivalry between correlated and independent cases) is contingent on an unrealistic presumption that people vulnerable to the chips illusion are knowledgeable enough about probability theory to be familiar with correlated as opposed to indepen-

---

6. Other illustrations are readily available. For example, in experiments intended to explore "anchor-and-adjust" effects Kahneman and Tversky (1984) found strong effects on group responses to World Almanac–type questions (such as "What is the population of Zaire?") from what number turned up under the pointer of a carnival-game wheel, where the wheel was spun in front of the group. Participants presumably did not see themselves as reaching judgments powerfully influenced by a transparently irrelevant event. Or in experiments exploring the propensity to see patterns even when none were present (Gardner 1985), subjects were eventually informed that the data they were looking at was completely random and were shown how the material was arbitrarily generated. The common response was for the subject to remain convinced that the pattern he thought he saw was, in fact, present.

dent events. Otherwise how could they be able to misread one situation for the other? But to suppose that the illusion would only affect someone familiar with the terms "independent" versus "correlated" would miss much of the point of this discussion. For not only has no such presumption been made, but no such presumption would make sense on the argument being developed. It is essential to see that the point is not in any way contingent on the person being *explicitly* familiar with the correlated versus independent distinction, any more than the explanation of difficulty in tying square knots is contingent on a person being familiar with the explicit idea of a granny knot.

If we want to make some explicit conjecture about what is cuing an intuition, we are (of course!) reduced to stating it in words. In this case an efficient way to state it in words is to use the language of probability theory. But the cognitive process that underlies the intuition is not contingent on knowing these words or with being familiar with the concepts in any other explicit way. Rather the terms are fundamental in probability theory because they characterize some pervasive aspect of physical experience. A reader totally innocent of probability theory has the physical experience that underlies that technical language. The distinction can firmly govern the intuitions of a person who has never heard of the labels.

Repeating that essential point: we live in a world where some events are correlated and others are not, and where a person will have much experience in getting punished for confusing correlated and independent cases (as, long ago as a child, he had much experience with the difficulty that arises if he ties shoelaces as a square knot rather than as a granny). Hence we all have acquired skills, but of course not unerring skills, in reacting one way to some events and another way to others, though without ordinarily acquiring any conscious awareness whatever of choosing between those ways—as you probably had no conscious awareness that you tie your shoelaces in a granny knot. Unless you are a rare person who was aware that tying shoelaces is tying a granny knot, you have in hand a very apt illustration of the point that learning, and later responding, need not involve any touch of conscious

awareness of exactly what you are doing or why you are doing it.[7]

3.10

But whatever merit there is to this analysis of simple puzzles, why
suppose they would apply to situations that are really important to
a person? Half of *Patterns* and most of *Paradigms* is devoted to
that question. I try to provide elaborate documentation of the
extent to which brilliant thinkers, devoting themselves for many
years to a matter, can be totally blind to features of the situation
that eventually come to seem wholly obvious. The examples I
review are mostly the same that Thomas Kuhn used in his famous
book on scientific revolutions, out of which the label "paradigm
shift" became a piece of everyone's thinking. Part of what Kuhn
showed was how these major episodes in the history of science
are marked by what he called *incommensurability*: a blindness of
at least one side of a controversy to the force of what the other
side was saying. This is an effect so striking that it became hard
for historians not to suppose that the people on the wrong side of
these paradigm shifts were just stupid or bigoted. But as Kuhn
showed, the people on the wrong side of these controversies were
not ordinarily more stupid or more bigoted or more of whatever
other unflattering label you are tempted to use, but nevertheless
somehow blind to arguments that the other side (and later generations, and sometimes even many of those on the losing side after

---

7. Many other illustrations are available. But some are so much a part of
normal routine that it is hard to notice them: e.g., our routine recognition of
faces, our production of grammatical sentences, and many other things we do
with no apparent effort but also with no capability to articulate how we do
all that. On the other hand, other examples are not so routine, but tend to
seem merely bizarre. There is a specialized task in the poultry business called
"chicken sexing," which turns on the need to decide which chicks are female
before any explicitly perceptible clue to that is available. People are trained to
be chicken sexers by giving feedback on when they guess right. Some people become very good at it, though they have not got a clue as to how they are doing
it. Successful dowsers, similarly, probably are responding to landscape cues they
cannot explicitly articulate: they are confident, rather, that the divining rod is directly responding to the presence of water.

a few years) saw as terrifically important. In *Patterns* and in much more detail in *Paradigms* I have tried to illustrate how that puzzle of incommensurability can be resolved. What that requires is that we attend carefully to the habits of mind that would govern intuitions prior to the paradigm shift and to circumstances that would allow some individuals to finally break free of those habits of mind.

The force of those habits of mind can be startling. If we ask what Copernicus knew beyond what any careful student would learn from Ptolemy (hence what every serious astronomer knew for the fourteen centuries from Ptolemy to Copernicus), the answer for all practical purposes is *nothing*. The Copernican argument, which is a very elegant but essentially simple argument, was available (logically) to any talented astronomer throughout all that time. But it lay on the table for fourteen hundred years and no one could see it. Similarly, the beginnings of probability theory, with its immediate and lucrative practical applications to gambling problems, was (logically) there for the picking by anyone with modest mathematical competence for at least two thousand years. But no one could see it, and then at first (c. 1655) only the three most brilliant mathematical minds active at the time caught it (*Paradigms*, chap. 6). And so on.

But if brilliant minds, deeply devoted to study of a matter, can miss what eventually becomes obvious, would it not be naive to suppose that you or I, when dealing with matters outside our particular special interests (i.e., matters in which we are not expert) could not also exhibit some blindness to the force of what logically are very good arguments?

To suppose good arguments are necessarily convincing unless we are dealing with someone who is stupid, corrupt, or lazy is like supposing that our perceptions are always reliable insights to the true nature of the world, rather than what our brains give us from cues that we have tacitly learned to use from our limited experience in the world. As perceptions are generally veridical (so that the world generally works at least "as if" those perceptions were right), so intuitions are generally reliable. But there is nothing automatic or guaranteed about that. When we are confronted with situations that are novel, impoverished of familiar cues, blurred, or otherwise odd or unfamiliar or difficult, intuition (like perception) will be vulnerable to illusion.

3.11

Now consider how this line of argument might be applied to the case of expert/lay conflicts of risk perception, where (following the argument of 2.7) what is critical is the *visceral* response to risk. By comparison, it is a minor and much less surprising point that there are also are differences between expert and lay estimates of the *statistical* features of risks. We want to pin down, as far as we can, what might account for experts feeling active concern about certain risks that are commonly neglected by the public, yet with intense public concern with other risks that experts see as trivial. Simple puzzles like the chips puzzle here give us an indication of how we might proceed.

Professional statisticians aside, the illusory prompting of intuitions about independent rather than correlated cases in the chips problem catches something over 90% of people even as sophisticated as the readers of this study are likely to be. But if we look at the very special set of readers who happen to be statisticians, then not many (though not, it is worth noting, none at all) are likely to have had difficulty.

The contrast between expert and lay responses here partly reflects the availability to the expert of formal technical apparatus not so fluently available to others, though the formal probability theory needed for this problem is so minimal that any literate person has that in hand. It is probably more important that a professional has a great deal of experience in dealing with formal probability from the work she does in applying and teaching statistics. A default propensity to feel events as independent (rather than correlated) would not so easily have its bizarre (in the "chips" context) effect on a person with that much experience. Similarly, a sailor is not likely to fall into the physical pattern that governs the way he ties shoelaces when he intends a square knot.

So the difference between lay and expert cognition on the chips or square knot problems (the statistician or sailor, as against you or me) is partly that we have different repertoires of patterns available to us that grow out of differences in experience. That entails differences in how easy or hard it is to cue one pattern (even if it is the default for all of us) as against a rival pattern (which is not the default), even when we all have both patterns in our repertoire.

This last condition requires special stress, since (tied to point 3) a person will easily be more confident that he knows too much to be vulnerable to some error than in fact he is. Logically, that is often right: logically, he knows enough so that he should not be vulnerable to the error, as logically a person setting out to tie a square knot knows how to do that. But the intuition may nevertheless be wholly wrong. When I note that professional statisticians are not usually vulnerable to the chips illusion, I mean that in a strict sense. As anyone who tries the problem on friends and colleagues will find, people who merely know statistics, but who are not professional statisticians, are more likely to recognize (prompted by Q4) that indeed they have contradicted themselves. But that people have some training and even that they use statistics in their work, by no means produces immunity to the illusion.

Parallel to the responses here to a simple puzzle, we will shortly start to explore an account of expert/lay conflicts of risk intuition in which the conflicting intuitions reflect different responses to common cues by people who differ in their experience with the situations in which some risk arises. In particular, I want to emphasize differences between people for whom the details of the situation are of a sort that is part of familiar experience *as against* individuals who are responding in a context that is outside their range of familiar experience.[8] The consequences will be those already mentioned in connection with the discussion of 1.3: given the same information, what the expert feels about a risk—the expert's visceral sense of how cautiously to treat this risk—may be radically different from the lay sense, as the usual statistician's response to Q1 and Q2 of the chips problem is wholly incompatible with the usual lay response.

8. A second essential distinction is between people who have some prior commitment (e.g., an ideological commitment) that would prompt them to find it easy to believe, or to find it hard to believe, what most experts on the matter at hand are claiming. But readers will already know that.

3.12

But in the face of such claims, we are prompted to ask (following Nisbett and Ross 1980) "If we're so dumb, how did we get to the moon?" The emphasis in this chapter on *illusory* intuitions partly reflects the special concern of the overall study with cases in which intuitions conflict, so that presumptively one side or the other is seeing things in an inappropriate way. But parallel with the prominent role played by perceptual illusions in studying how vision works, illusions of judgment can yield insight into how *effective* cognitive processes work. For we do not expect that evolution would provide us with special processes which generate illusions. That makes no sense at all. Rather, we expect that when we find illusory responses we are seeing normal cognitive processes that, under the peculiar conditions at hand, generate illusions.

Hence, the argument of this chapter has turned on how, in certain circumstances, ordinarily effective functioning of habits of mind can yield illusory judgments. That is not an argument that intuition is ordinarily unreliable or that careful reasoning is always ineffective in challenging faulty intuitions. But cognitive illusions do occur, can be very stubborn, and it is highly implausible to suppose that illusions occur only when they can have no bad consequences. So we have good reason to want to understand the conditions under which problems about such matters can arise and especially when they can be particularly severe. The conflict between expert and lay intuition is not absolute, in the sense that every expert and every lay person will respond the same, just as not every lay reader is caught by the chips illusion and not every statistician is immune. But a marked tendency of that sort is easy to observe. And, as we will have occasion to discuss, in a social situation there is a certain contagion to such tendencies.

The Risk Matrix

4

We want to consider an account of expert/lay conflicts that runs parallel to the account of cognitive puzzles in chapter 3. The argument is that in contexts out of the range of familiar experience—artificially narrow, as with word puzzles, or extremely broad, as with many questions of social policy—ordinarily effective cognitive processes can lead to confident intuitions that eventually come to look plainly wrong. And since we have seen that self-contradictory intuitions can be prompted within a single individual dealing with a simple situation, it could hardly fail to be the case that, in dealing with some complicated social context, conflicting intuitions will occur across different individuals with different experience. And, again parallel to the situation with puzzles, sometimes one view of the matter turns out to have a reasonably clear claim to making more sense than its rival.[1]

As illustrated in chapter 3, we can expect, for an individual operating outside the range of his normal experience, a much stronger role for defaults and other cognitive features cruder than we would see in some context where a person has acquired a repertoire of re-

---

1. That is not always the case even for simple laboratory problems. A much-studied phenomenon is preference reversals with respect to certain kinds of gambles (Slovic and Lichtenstein 1983). A large fraction of intelligent subjects give conflicting reports, but it is not a matter of one response is right and the other wrong. The subject just seems to disagree with himself, contingent on how the question is asked.

sponses well tuned by feedback from experience. We want to see how, under some conditions, this would yield cases where the characteristic lay sense of risk is high, but the expert sense is mild. But under changed conditions, the same account (if it is plausibly adequate) must account for the converse cases, since these also occur, where experts are worried but laymen are not. And of course the account must under yet different conditions work for the most common cases, in which there is no conspicuous conflict.

But if there is no conflict, there is no occasion for intense interest. And if the characteristic lay response to a risk is to worry markedly less than experts think prudent, then the characteristic political response is to relax efforts to enforce the expert view. Indeed, since lay indifference coupled with expert concern typically involves some matter of individual choice, the case for government intervention is weakened not only by widespread indifference or opposition but also because any particular individual is mainly harming only herself if the experts are right but ignored. So expert concern coupled with lay indifference usually would not lead to sharp controversy. It is the reverse case that leads to intense public concern.

Most of the early discussion of expert/lay conflicts in fact was discussion among experts dismayed at lay indifference to expert warnings about such matters as cigarette smoking, seat belt wearing, and property insurance for people living on floodplains (Kunreuther 1978). But the environmental issues that came to prominence in the 1970s are social in character. If I am concerned that smoking is dangerous for smokers, I can stop smoking, but if I am concerned that second-hand smoke is dangerous, I cannot leave the environment. These cases have been common since the 1970s, but we can point to cases that have a contemporary flavor dating back to the 1950s— which is before any marked loss of trust in government and experts had occurred—with fluoridation as a salient example.

4.2

It is convenient to settle on some language that captures the rival senses of context that might account for conflicts of expert/lay intuition. But since we are looking for contexts that would be familiar from experience in the world, we can also expect that in one way or another we will find language already in use that corresponds to these situations, as with the probability language that was available for the chips problem. Since the argument is that what governs responses is not open to introspection (3.6), why should we expect explicit language? Because we attach labels (give names) to familiar things, but in particular instances we often recognize those things only subliminally. So suppose we could have named the context if we consciously noticed it. That does not imply that we only respond if it is consciously noticed. Nor does it imply that we will have some compact language at hand: it may be something looser, for example, something like the proverbs I am about to use here.

The kind of language I want to invoke uses bits of proverbial wisdom. In one sort of context, some such language (some proverb) will seem salient; in a contrasting context another comes to mind. For proverbs in fact tend to come in pairs that capture rival insights. A notable feature of these pairings is that we are ordinarily wholly unconscious of the rival pairs, so that it comes as a bit of a shock to a person the first time she notices that proverbs so typically do come in rival pairs. Instead of seeing rival proverbs, we experience them with a gestalt sort of response, as with the scenarios of chapter 3. We see things one way or the other, and when we are seeing things this way, we do not see the other way at all, though in other contexts—perhaps even another context very nearby—we can recognize that "other" right away. When intuitions are being governed by a particular sense of context, so that one side of a proverbial pair is salient, the competing sense of things will be inhibited. When we see "too many cooks spoil the broth" in a context, we do not notice that at other times we see "many hands make light work."

Sometimes we are in a state of tension, so that we find ourselves alternating between one way of seeing and its rival. For a gestalt drawing (the duck/rabbit, e.g., in fig. 4.1), this is the usual situation. But when we need to act—even if only in the sense of voting or deciding what side to root for—such a state will rarely be sustained for long. Rather, when a pair of gestalts are rivals (i.e., are mutually incompatible), we will have a tendency to lock

Fig. 4.1—Koehler's duck/rabbit

in on one or the other, as a creature in a situation where it might fight or might flee will lock in on one or the other, not waver between the two. Or we lose interest and our attention goes elsewhere.

For risk assessments, the pair of proverbial contexts I want to propose is "better safe than sorry" versus "waste not, want not." The first signals caution; the second signals getting on with things, with attention to the costs of delay and foregone opportunities.

Such intuitions are commonplace: a person repeatedly encounters situations that call for moving more cautiously or for getting on with things. So sometimes "better safe than sorry" seems appropriate, and other times it is "waste not, want not," or "nothing ventured, nothing gained" or "he who hesitates is lost." That the world presents a creature with many situations that require prompt choice between rival responses has important Darwinian consequences. As with the competing "independent" and "correlated" possibilities of chapter 3, a creature that survives learns to respond to situations that require choice between rivals, because the world is full of such situations, whether the creature involved can use language or not.

From statistical usage, an alternative, equivalent, characterization of the rival gestalts contrasts a focus on "errors of the first kind" (Type I errors, or the risk of taking an affirmative step which is a mistake) versus "errors of the second kind" (Type II errors, or the risk of failing to take a step that would be good). But whether trained in statistics or not, a person would somehow have a sense of such ubiquitous conflicts. That must shape the endemic propensities that govern how we think, as discussed in some detail in *Paradigms* (esp. chap. 10). Yet another, still equivalent characterization exploited in *Patterns* is the tension between "jumping to soon" and "hesitating too long."

What I want to propose is an account of various ways—but one of these in particular seems critical for expert/lay controversies—in which a person might come to be in one of the polarized states, seeing only "better safe than sorry" and blind to "waste not, want not." That is, this person sees only the risk of errors of the first kind and is somehow blind to errors of the second kind; seeing only the risk of jumping too soon and being oblivious of the risks of hesitating too long. For, as a matter of logic, everyone will agree that we need to attend to both sides of all of these pairings.

4.3

If there were nothing to gain from taking some risk, then of course we would avoid it. And unless some danger is averted we would not incur the cost or inconvenience of taking precautions. But if we see both risk averted and costs incurred by taking precautions, then we look for some reasonable balance between caution and boldness. That is not contingent on attitudes toward formal cost/benefit analysis. Nothing has been said to limit costs incurred by taking precautions, or benefits foregone by not doing so, to aspects that are ordinarily included in economic analyses. Rather, the sort of judgment at issue merely reflects a rudimentary tendency to economize on resources and enhance values. That could hardly fail to be somehow entrenched in the habits of any species that survives in a Darwinian world.

Consequently, parallel to the pairs of competing scenarios discussed in connection with puzzles in chapter 3, we could expect that we all have some propensity to see risks in terms of opportunities that might be available if we accept a risk and of danger that might be averted if we take precautions. This means that, in contrast to the rival scenarios used to define gestalts for the cognitive illusions, here we will need two dimensions to define the

Fig. 4.2—The risk matrix

contexts: one responding to the danger we see in accepting a risk, the other to the costs (opportunities foregone) of taking precautions to avert the risk.

In figure 4.2, costs of precautions are on the vertical axis, and the danger of doing without those precautions is on the horizontal axis. Both are to be understood in a wide and intuitive sense, by no means necessarily limited to costs or risks measurable in dollars, even in the case of a person inclined to assess things that way. On the other hand it is whatever costs (and negative costs: benefits) that are *on-screen* for a person, not what an expert might assert ought to be considered, that is relevant here. Commonly, the costs of avoiding one risk include the costs of accepting some other risk, as will be conspicuous in the example of the saccharin case and others to be reviewed in chapter 6. But a person might not be sensitive to all risks and costs in a situation, but focused on one kind of risk or one kind of cost.

Setting out the possible combinations with respect to whether costs of averting a risk and costs of accepting that risk are getting attention, we get the 2 × 2 *risk matrix* of figure 4.2.

In the lower-right cell 4, neither the cost of (marginal) precautions nor the (marginal) risk of not taking precautions are salient. The whole situation is *off-screen*. We may know it is there, but it is not a focus of active attention. But in cell 1, a person is alert to the situation, with the costs that go with the risk on-screen, and also the costs of taking precautions against that risk. What characterizes this context is *fungibility:* we are in a situation where we can see a need to consider, and usually a need to some-how balance or trade off, the advantages of caution against the advantages of boldness. Usually we must give up values on one dimension to secure value on the other.[2]

For the remaining two cells (2 and 3), either costs or benefits, but not both, are on-screen. Response is dominated by one side or the other, and (here calling on the discussion of rival proverbs in 4.2) intuition is then governed in a one-sided way either by "waste not" intuitions (cell 3) or "better safe" intuitions (cell 2).

When the response is active but contested (cell 1), we can expect an attempt to reach some balanced judgment, to see both sides of an argument, to seek expert advice, and so on. If we reach a balance, so that nothing much seems to be gained by shifting a bit either toward more boldness or more caution, we will eventu-ally slip to cell 4. For if we have no clear preference about which way to move, we come to feel comfortable with, or at least re-signed to, where we are.

Movement through the risk matrix would ordinarily both start and end in cell 4. Suppose attention was first drawn to a source of danger, putting the person into cell 2. But then the individual might notice that, as usually is the case, reducing that risk also has costs, a realization that moves the context to cell 1. But in cell 1, we can expect a person, having considered that bal-ance, to reach a sense that favors either more caution (cell 2) or more boldness (cell 3), though the latter might just be a matter of judging that the previously unnoticed hazard is not worth wor-rying about after all. A person tries to reach a situation where a sense of balance between boldness and caution is felt; this is fol-lowed by a waning of interest, when things seems to be about right, and a return to cell 4, where the situation is again off-screen.

Within this process, a person might waver, seeing things as requiring more caution at one moment, but then noticing other features that seem to argue for more boldness. Adjustments one

2. What about a case where a dominant move, one at once safer and bolder, is available? Whatever cell we were in, such a move should look attrac-tive, though for someone strongly polarized in cell 2, we might anticipate a skep-tical (but perhaps not irreversibly so) reaction to a claim that some move that is bolder is in fact also safer.

way might come to seem to be going too far, so that a contrary adjustment is needed. And it may be that the balance that feels right is hard to see, so that only after a good deal of deliberation and seeking advice that we can choose what feels better. For keep in mind that it is the visceral feeling of risk, not a statistical calculation that ultimately will be controlling (2.11). But the general tendency for persons confronting cases in the normal range of experience—for cases that can be placed somewhere in the middle zone of figure 1.1—is for a trend of the sort I just sketched: starting in cell 4 (off-screen); coming to attention, probably at first focused on a newly noticed danger (favoring cell 2) or opportunity (cell 3); then noticing that the opportunity entails risk, or that worrying only about the risk entails forgoing opportunities—so we reach cell 1; and then the process as sketched back to cell 4.

But sometimes a person cannot make a comfortable adjustment. Then things take a more marked course, where costs of accepting the risk, or where costs of precautions imposed, dominate, and in a stronger sense than the transient, or tactical, sense of the previous paragraph. Then we get cases where a person is effectively locked into cell 2 or cell 3, as a person responding to the chips puzzle commonly seems locked into seeing the 1/2 response as beyond doubt. Either danger is above the threshold, but the costs of precautions are not, or the reverse. Then there is no sense of something to balance, trade off, or have doubts about. What is needed is action to push things to a more reasonable situation. If a contrary challenge comes, it prompts debate mode (someone is making a perverse argument and you want to show that), not discussion mode (someone is making an argument that might be right, and you want to consider that). It then takes something particularly striking or persistent to pull the off-screen component of the balance back on-screen.

The cases that concern us are just these locked-in cases, in which a person is not in one of the polarized cells in only a tactical sense. In that merely tactical situation, even though just now one side seems paramount, a person remains open to considering both the risks in accepting a danger and the costs of avoiding it. But in the polarized situation, the person feels locked into what she knows is wrong. Attention is locked onto the aspect (danger or opportunity) linked to that. We can observe that this locked-in situation is not in fact something rarely seen and likely to affect only an atypical person or a normal person only in a very atypical situation. Rather, it is the normal, or usual, response in a class of common situations.

4.4

Suppose two individuals with differing experience face the same risk. Then one person might see "waste not" while another sees "better safe." That would be no more surprising (for individuals with different experience in the world—e.g., one is an expert in this matter and the other is not) than that one of a pair of animals would choose to fight in a context where another would choose to flee. But since human beings, unlike other creatures, often face choices in which one decision must be made for all (we make social choices, not just individual choices), a social difficulty appears. There is no conceptual difficulty in understanding how such a situation could arise. If the choice is social, I am stuck with your choice unless you are stuck with mine. We face the "locked-in" situation mentioned a moment ago, where a person may have no way to respond to visceral discomfort, since by himself he cannot move (or can only move at a cost that seems unreasonable) to a more comfortable setting. And as social creatures we may be called upon to vote or otherwise influence a choice that involves other people. So even when we are not personally at risk, we may experience visceral discomfort at the situation being imposed on other people.

Expert/lay conflicts can arise for both situations—individual or social choice—because in any context at all, there will be cues that are subtle or complicated or otherwise difficult to use for an inexperienced person, but are familiar and automatically significant for the experienced person. Just what we mean by *expert* is that a person has a lot of experience on some matter. A pair of individuals—one expert, the other not—may respond differently to a common set of cues. Subjectively they see different situations, though both agree that objectively it is the same situation. But why that is happening cannot be directly observed, since (repeating that essential point once more) the context patterns that govern our intuitions, and the way we use (or discard, or never notice) cues are inaccessible to direct perception (3.6).

For social issues, if the predominant lay response is firmly in cell 2, political realities will easily push precautions beyond what seems sensible to expert judgment; this puts experts in cell 3. But, given pragmatic constraints on how much expense and inconvenience can be managed, those precautions might still fall short of what will move lay judgment out of cell 2. So we get a polarized situation, where a one-sided sense of the situation for some set of individuals (mainly lay) eventually leads to an opposing gestalt for another class of individuals (mainly expert). People with one kind of experience in the world find themselves in cell 2, while others—equally intelligent, well-motivated, and so on—find themselves in cell 3. Looking across that axis from either side, a person sees people whose judgment is perverse, narrow, untrustworthy, and so on.

Such reactions can occur even when only individual choices are at issue: I think it is pathetic you are so timid, you think it is crazy that I am so reckless. But the conflicts will be most intense in the context of social choice, since both sides are locked into living with a common choice, which looks too timid to one side and too reckless to the other. And it is also in the context of social judgments that social knowledge (what "everyone knows") is particularly likely to have taken hold, and further constrain the possibility of mere reasoning to shift intuitions.

But the story, to this point, can be read in a way that still fits comfortably with the "usual story" discussed in chapter 2. In that usual story, a central aspect of the polarized situation is that expert intuition is focused narrowly on statistical expectations of damage, but lay intuition is shaped by many other dimensions, as illustrated by Covello's list (fig. 2.1). So although it is indeed the contrasting experience of experts and nonexperts that accounts for the conflicting intuitions, what is most important about that contrast is that the experience of experts makes them focus narrowly on what they have been trained to attend to. Lay intuition is no doubt not so fluent at responding to subtle information about

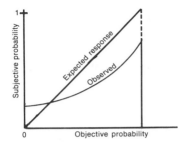

Fig. 4.3—Relation of subjective judgments of probability to actual probabilities.

statistical risks. But, in the usual story, what is accounting for the stubborn conflicts is less what experts *see* that other people *miss*, but what ordinary people *feel* about risk that experts *neglect*. But I have from the start made it clear that I want to present a very different view, turning on a blunter dichotomy.

4.5

Can we really suppose that this complex problem turns in a dichotomously simple way on the working of thresholds? For logically it is certainly the case that we are almost always in a state of fungibility (cell 1), where there is some trade-off between costs of accepting and costs of averting a risk. So why give such prominence to the "locked-in" (as opposed to merely tactical) cases of cell 2 or cell 3 intuitions? In fact, though, there are many lines of evidence suggesting that, whatever logic might imply, cognitively, human beings very often see their world in a dichotomous way. In responding to risks, this propensity is strong when quantities associated with costs or benefits or probabilities are hard to see.

Example 1.—Figure 4.3 shows the characteristic pattern of responses revealed by a wide range of experiments and field studies. On the vertical axis is a person's subjective sense of probability, and on the horizontal axis is the objective situation, as shown, for example, by data for a long series of events. If subjective probability were an unbiased estimate of objective probability, and we average across many such estimates, we should approximate the figure's diagonal line. The value on the vertical (subjective) axis would be the same as on the horizontal (objective) axis.

But the consistent result is a curve of the sort shown, which lies above the diagonal close to probability $P = 0$; below the diagonal thereafter. What is relevant here is that there must be a jump at each end of the curve. A person's subjective sense of probability does not map straightforwardly onto an objective assessment. Instead, intuitions about probability close to $P = 1$ or $P = 0$ take on a dichotomous character. The outcome is treated either as if it were $P = 0$ (or $P = 1$ at the other tail), hence certain (no risk). Or it is treated as definitely risky, and hence cannot reasonably be treated as negligible.

But the cases of sharp expert/lay conflicts of intuition are in fact usually cases where the statistical risks to an individual are very small—indeed, microscopic. No one could sensibly argue that such subtle risks are always to be treated as negligible. But it is worth some attention when we observe that risks that are actually very small prompt intense lay concern even though the same individuals see worrying about otherwise very comparable risks—but in another context—as neurotic or stupid. And in yet other contexts, risks actuarially far larger than those prompting intense concern here may be treated as obviously negligible.

So although it is only the tails of the Kahneman and Tversky curve that show the dichotomous on-screen/off-screen character stressed here, our concern is precisely about what happens at those tails. In the language I have been using, risks that are statistically microscopic somehow come to prompt very substantial visceral perceptions of risk, while much larger risks are ignored as negligible.

Example 2.—A second indication of the propensity to dichotomous response comes from studies of insurance. Both in experiments and in the field, people reveal a bimodal response in their willingness to pay for insurance against small risk (see fig. 4.4). Again, a person either treats the risk as negligible, hence is willing to pay little or nothing for insurance: or the risk is treated as significant, in which case the person is willing to pay substantially more than the expected value of the risk. Somehow, something

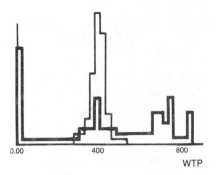

WTP

Fig. 4.4—For a .4 risk of losing $1,000 (light lines) response approximates a normal curve; it is centered on the expected value. For .01 risk of losing $40,000 (dark lines) normal response fades and polar responses become obvious. For cases relevant here, which would involve risks much smaller than .01 but possible losses valued more than an amount of $40,000 (i.e., getting cancer), normal response almost disappears and polarized responses become dominant. (Coursey et al. 1993.)

pushes or fails to push a risk above a threshold, in patterns that often seem to leave the statistically calculable risk as irrelevant.

Example 3.—A similar interpretation fits a third well-documented feature of reactions to risk. As Kahneman and Tversky and others have found, willingness to pay for a very small increment to safety is usually near zero *except* when that small increment is the last increment.[3] So a person will often be interested in an opportunity to change a risk from some very small value to zero, but probably not for that same increment if the resulting value is much greater than zero. Consequently, there is a reliably big ratio between willingness to pay for a change from, say, .001 to zero risk, compared to a change of the risk from .482 to .481. In terms of the argument here, that would reflect a tendency to see .001 as negligible when the context anchors attention on .482, but significant when the anchor is zero. This is usually referred to as the "certainty" effect, since it is most stark for removing the very last element of risk.

Such responses can be made sense of—that is, made plausible

3. Kasperson and Stallen (1991) survey this and related results.

though not necessarily reasonable—if we take into account bundling effects (introduced in the next section), which shape what a person perceives as a zero-risk situation. For although logically we all know that there is risk everywhere and always, so that "zero risk" is never a choice truly available, our visceral perception of risk is very different. We see large variations with respect to what it takes to prompt a jump from a risk's being treated as negligible to being seen as significant, or how large an increment of risk might be and still be seen as negligible. This routinely occurs within the same individual, as he responds to different contexts. And these effects are often so stark as to seem simply bizarre if presented out of context.

Example 4.—More generally, the peculiar history of quantitative probability (*Paradigms*, chap. 6) complements the threshold argument under review here. The puzzle is to account for the long delay of the emergence of what are now commonplace ideas about probability as a number. Quantitative probability first appeared in the 1650s, after two thousand years during which everything needed for its discovery was available but never fully exploited. Apparently there is something highly artificial—hence hard for a person to do until he has seen someone else do it—about attaching a number to a probability. It is a reasonable conjecture—and in fact a strong one, given the other evidence already mentioned— that although now everyone is familiar with simple notions of probability as a number, some residue of the earlier difficulty remains. This would most easily appear for the sorts of probabilities we would have little practical occasion to think much about: that is, for probabilities close to zero or one.

So probability in the corners of the Kahneman and Tversky diagram tends to be treated as either indistinguishable from zero (negligible), or as clearly substantial—as if the less probable outcome had a chance somewhere close to 50/50, though the person logically knows perfectly well that it is more than the first and less than the second. If you are tempted to pass someone going uphill on a country road, the probability is small that an oncoming car will appear on the crest just as you are in the midst of that maneuver. But the probability *feels* like 50/50—not like something remote but nevertheless a chance not worth taking.

In sum, then, evidence of several kinds (from experiments, from field studies, and indirectly from the history of probability) suggests threshold effects with respect to subtle risks. We tend to see such risks in an "either/or" way: either the risk is significant (so it gets treated as if it were something like 50/50) or it is negligible (so it gets treated as if it were zero). And this is the kind of cognitive response consistent with what I am calling the strong interpretation of the risk matrix. The danger is seen as either too small to worry about or as too big to accept, with no middle ground.

The consequences of that dichotomizing propensity would vary from person to person and context to context. It does not at all follow that if one person *feels* a particular *statistical* risk as negligible, then another person will share that *visceral* response even if he believes the probability claim. Nor does it follow that the same person, but now in a different context, will see the same danger with an even smaller probability as also negligible. We now start to explore how that works and what it takes to move a person to where he can at least see the dichotomy both ways. We ought to be able to do that, as we can all see the familiar gestalt drawing (fig. 4.1) both ways, though at a moment when we see the duck we cannot see the rabbit and a moment later when we see the rabbit we cannot see the duck.

4.6

Parallel either/or effects are conspicuous on other matters, of which one aspect is important enough here to warrant separate treatment. The first principle of toxicology is that "the poison is the dose." Such language is prominent in the opening pages of any text on toxicology, just as "keep your eye on the ball" is a standard caution in many sports. The emphasis and immediate attention given the advice reflects (for both cases) the difficulty novices have in absorbing the advice. Here, as in many other skilled activities, mentors need to drill novices to overcome their natural tendencies.

For the ball game, it is easy to see why natural habits may be bad habits. In almost all contexts, seeing what will happen next requires a different focus of attention from seeing what is happening just now, so that in any interval we do best to pay some attention (and hence prepare for) what will happen next. The peculiarly structured characteristics of dealing with a ball in a game creates situations in which an artificially exclusive focus on just the ball is best. But toxicology (in particular the sorts of subtle toxic risk of special concern here) is likewise about situations where good performance would go against the general run of experience. For in ordinary experience it would hardly ever to make sense to hesitate (invest time in study, etc.) over a choice whose consequences one way are almost certain to be indistinguishable from consequences the other way. Instead, as with probability, we tend to binary (polarized, dichotomized) intuitions. The substance is either a poison or it is not. The risk is significant (treat it as roughly $P = .5$) or negligible (treat it as $P = 0$). Similar overly stark tendencies can be observed with respect to friend/foe, good/bad, and even to arbitrary pairs, like "ping" versus "pong."[4]

For substances, the first cut is that either they are associated with bad outcomes (they are dangerous, so stay away from them) or they are not. Only if there is a noticeable cost to the binary reaction—there are things we want to do but cannot—do we develop more refined intuitions about due care but not excessive care. Even then, through "bundling" (discussed next) we eventually return to the binary propensity. But now we may feel the danger as negligible, where earlier it was, with equal clarity, too dangerous to ignore, as will be illustrated in chapter 6.

4.7

We want an account specific enough to be put to work, one that needs to tell us something of what might push the sense of some costs above threshold and leave others below threshold. We want

---

4. Gombrich gave people arbitrary lists, and asked them to classify items as "ping" or "pong." People exhibited no difficulty carrying out this nonsense task: Mozart—ping; Beethoven—pong, etc. See Perkins (1981).

to be able to say how a person comes to be in one rather than another cell of the risk matrix and consequently also how a person sometimes moves (or better: is moved, since this is not something we can just decide to do) from one cell to another. But all such responses must be made against a baseline sense of a usual situation, which will already include various features intended to control risk. This is easiest to see for activities like parachute jumping or scuba diving, which in the absence of considerable precaution would simply be methods of committing suicide. Parachute jumpers and scuba divers are not struck with terror at the thought of doing what they go to a lot of expense and trouble to do. Nor are they immune to normal responses of visceral risk. Indeed, such responses can be easily prompted—but not by a routine jump or dive. That routine has already bundled into it, as just stressed, a set of usual precautions that provide (for jumpers and divers, not necessarily for you and me) a comfortable balance between risk and satisfaction.

But this *bundling* point holds in general, not just in obvious cases like parachute jumping and scuba diving. My sense of the riskiness of a shipment of radioactive waste cannot be some abstract sense of the material somehow moving from place to place independent of any at least implicit notion of what that would involve. Even if nothing is specified, I will have a tacit sense— perhaps realistic, perhaps wholly unrealistic—of what hazards are involved. And the same for waste dumps, asbestos in schools, microwave radiation, bungee jumping, and anything else that might arise as a possible source of risk. So when we consider threshold responses for costs of accepting versus avoiding risk in some context, we need to do so with close concern for what the baseline might be of the customary or habitual costs, benefits, risks, safeguards, and so on, that are part of a person's sense of what is normal. This will naturally vary with each person, since what is perceived, readily assimilated, and so on, must be highly contingent on that person's particular situation and experience.

In *Patterns*, I used the label "bundling" in another sense: that of coming to see a set of related patterns (such as familiar logical patterns used as steps of an elaborate argument) in a single all-at-once way. But the two senses each reflect the cognitive tendency to build packages out of what initially were cognitively separate items.

4.8

The "bundled" (baseline) situation might not be anywhere near equilibrium, where costs of accepting versus costs of avoiding some risk seem in balance. Rather, most of what this study is about turns on just those cases where a person is polarized in one of the out-of-equilibrium cells (2 or 3; see fig. 4.2), so that the baseline situation itself is one that is uncomfortable. A person will not remain in constant distress about this, as will be discussed in a more detailed way when we consider habituation and sensitization (4.9). The out-of-equilibrium situation will fade from attention, but when attention is once again aroused (perhaps only momentarily by some transient anomaly) the person is immediately aware that he is out of equilibrium. A pair of contrasting examples that will be considered among others in chapter 6 will be the case of dioxin, which "everyone knows" is a dangerous carcinogen, as against aflatoxin, which many people will not recognize and few people are readily distressed about; still, human exposure to aflatoxin is vastly more substantial, and the evidence for its carcinogenic significance (at levels actually likely to be encountered in the environment) is far stronger. By and large, we are *sensitized* to dioxin and *habituated* to aflatoxin. We will want to consider how that happens. But the more general point is that discussion must start from some sense of a situation, not from a ground-up total catalog of risk. Even in an entirely novel situation, a person can never be without some background sense of circumstances. Even a novel situation is apprehended (often only tacitly) by sensing it as akin to something familiar; accordingly, it will carry along the background bundling of that situation. So although the response may involve some *adjustment* from that (often only tacit) "looks like" prompting, it is *anchored* there and will not be easily moved far away.

Occasionally debate or reflection or striking new bits of information can succeed in unbundling some component of this background. But that is not something that occurs easily, or whenever it would be reasonable, or in any other sense that makes it automatic or routine. Nor is it something a person can ordinarily consciously notice. A situation comes to look different to me. I may even be surprised by that. But I rarely can say just what it is that has changed, any more than mere introspection will tell a person why the duck now looks like a rabbit. It is a critical point (as already urged) that this absence of transparency (we cannot just look and see what is guiding intuition) will be especially significant for just those cases where intuition is being guided not by direct experience, but by more remote "looks like" or "everyone knows" intuitions. Then a merely logical critique of misperceptions about the case at hand will tend to be seen as irrelevant, or as a trick of some sort, or as merely unconvincing. This parallels (yet again) what we can observe for artificial illusions of judgment, as discussed in chapter 3.

4.9

What activates or disengages attention is one of the oldest topics of experimental psychology. In creatures all along the scale from worms to people, what is always found and what makes Darwinian sense is that alerts are triggered—a matter gets above threshold—by things that are out of line with usual experience, plus things that are familiar but which experience has taught us to handle with care. So familiar things are ordinarily handled "in the usual way," often with no conscious awareness at all for substantial periods. Disturbances that would put an inexperienced person on alert, and require concentrated effort, are ignored or tacitly adjusted for by someone with much experience. Of course, a person is not unconscious during such periods. Rather, she is conscious of other things, which is the point of the propensity to economize on attention. We can only explicitly *attend* to one thing at a time. But there is obviously an enormous advantage to being able to *do* more than one thing at a time (drive and carry on a conversation, etc.).

So when things are going along "in the usual way," control is largely unconscious with only transient explicit attention. But we are also put on alert by many familiar situations (Goffman [1959] gives the homely example of unpacking eggs) because we have learned to take care. This is an example of the case of sensitization. Finally, even with habituation, something that jars our sense of things going along in the usual way will push our attention above threshold. And since we talk to each other, attention is also capable of contagion. In particular, people who have been sensitized to some issue (rather than habituated), and so are easily prompted to give it attention, will want to persuade the rest of us that we ought to be on alert too.

On any live issue whatever (since otherwise we would not have a live issue), there will be some segment in the community easily prompted to active concern. Ordinarily, on any particular issue, this will only be some small minority, but that minority will naturally be motivated to provoke concern in the rest of us. And when their efforts succeed with enough of the rest of us to provide some socially critical mass, the political and social system as a whole will respond.

Years later—sometimes only many years later—there is likely to be a wide consensus about whether that provocation was a good thing (so we recall the provocateurs as heroes, as with antislavery activists, Dreyfusards, Churchill and his few allies in the years of appeasement, etc.), and other times the provocation will come to seem perverse (as with Salem's witch-hunts and McCarthy's Communist hunts). Finally, of course, and most common, there will be simply a recollection that some matter came to be high on the public agenda for a time with a generally good or somewhat unsatisfactory resolution or perhaps with just an eventual fading away, but with no conspicuous sense of either heroes or villains. But overall, there is an essential social contagion aspect to our topic.

Now on private matters everyone recognizes that unreasonable responses can occur. A person might come to regard his own fear—of flying, or heights, or whatever—as irrational. From firsthand experience, everyone is aware that we sometimes have habitual responses (e.g., with respect to overeating) that we would like to change but have not managed to change. When the responses are about social choices, as in the cases that concern us here, further complications arise. A person may not be easily motivated by a sense of responsibility to doubt his response since his individual response is such a small part of the ultimate social choice. Indeed the contrary motivation is more common: a person is inhibited from questioning intuitions that seem to be socially approved.

All this applies to human judgment in general, to expert judgment as much as to lay judgment. But a person's visceral responses, whatever they may be, can only be responses to cues entrenched by experience. And (yet again) what is familiar to an expert—for example, calculations with microscopic probabilities— may seem irrelevant or meaningless to a lay person.

4.10

Here is where we stand: drawing on what can be learned from very simple cases (from artificial puzzles), we now have on the table a particular hypothesis, summarized by a strong interpretation of the risk matrix. Costs of accepting a risk, or costs of avoiding that risk, or both, might be either on-screen or off-screen, yielding intuitions dominated by "waste not" *or* "better safe," or (for the "both" case) yielding tension between the two— the case of fungibility. The strong interpretation applies to cases where a person is locked into one of the off-diagonal cells (cell 2 or 3). The conjecture we are beginning to explore is that the expert/lay conflicts of intuition of concern here occur when the usual lay response is in fact polarized in this strong sense.

When a weak interpretation holds, a person would be in one of the polarized cells only in a tactical sense. She might be seeing danger as dominating opportunity (cell 2), or the reverse (cell 3), but in a way that never closes off awareness that a balance needs to be struck between boldness and caution. For that tactical case, a person does not lose access to the contrary perspective. Rather, her cognitive situation is analogous to the way a person seeing the rabbit (in the familiar gestalt drawing) is aware that the duck is also a reasonable interpretation, even though at the moment she can only see the rabbit.

But for the strong case, a person just does not see costs to caution (for cell 2) or costs to boldness (for cell 3), though sooner or later, if things are pushed to an ever more extreme position, sufficient starkness will push the missed dimension on-screen. Until we are confronted with cues stark enough to tie readily to our concrete experience in the world, we easily miss one side or both (danger or opportunity) when dealing with a situation outside our normal range of experience.

So long as *both* danger and opportunity are off-screen, the matter is left to others (bureaucrats, or experts, or politicians, or even persons unknown). If both danger and opportunity are on-screen, then we have fungibility and, consequently, an awakened interest in whether the matter is being well handled. Ordinarily that means relying on the judgment of people with particular experience with the matter. But outside the range of normal experience, cases in which we are alerted to both danger and opportunity are least likely, cases in which we are alerted to neither are most common, and the cases in which many people are alerted to either danger *or* opportunity, but not both, although rarer than the "neither" case (cell 4), are just those that generate controversy and hence come to wide attention.

Logically, that is not how things should work. But on the evidence I have been sketching, it is not so surprising that cognitively things should in fact work out that way. Then we will be easily prompted to very firm intuitions that treat one side or the other as negligible even when that is not at all plausible as an assessment of what is actually known about the situation. We get cases in which experts are worried but ordinary people are hard to persuade (as for some years held for seat belts, cigarette smoking, etc.). And we get converse cases in which ordinary citizens are very worried about something experts see as not very serious. I would argue that these polarized expert/lay cases for environmental risks should ultimately be seen as a special category of the more general phenomenon of social and political polarization. But developing that goes well beyond what this study can attempt to cover.[5]

5. I sketch an account of political polarization in *Patterns* (chap. 12).

4.11

Another element can now be noticed that will aggravate the situation. The least controversial single point in the psychology of judgment is that we usually pay more attention to risk of loss than to chances of gain (Tversky and Kahneman 1981). If we imagine looking at losses and gains of equal size through binoculars, losses tend to be seen magnified relative to gains; gains (relative to losses) tend to be seen as if we had turned the binoculars around. Other things equal, when confronted with a risk that is at least momentarily salient (so relative to where we see ourselves now, we are confronted with a possible loss), the default gestalt to expect is the cautious (i.e., loss-sensitive) "better safe than sorry" cell 2 of the risk matrix, not the perception of fungibility of cell 1. And the least likely response is the bold (i.e., gains-sensitive) "waste not, want not" gestalt of cell 3.[6]

But of course this tendency could not be absolute, either in the sense that a person would always be prompted to "better safe than sorry" (the opportunity for gain may be striking) or in the sense that the "better safe than sorry" response (if it is there) must be the strongly polarized "better safe" state rather than the merely tactical sense in which there is only a presumption (open to discussion) in that direction.

On its face, in fact, the risk matrix is indeterminate as to what

6. Suppose a propensity exists that favors one side of a pair of rival gestalts (context patterns) parallel to the "independent vs. correlated" pair that was so prominent in chap. 3. Then, on the argument so far, and indeed on commonsense grounds, the force of that propensity would grow as we move toward contexts that are cognitively difficult (complicated, noisy, ambiguous, far from familiar experience). Using once again the homely physical example of the square knot, if a person tends to tie a granny when a square knot is intended—and does so even when fully alert and making an effort to do it right this time— that effect is going to be all the more marked when the attempt is made in haste, when tired, when under stress, etc. This is the point laid out at the start of this study and elaborated in the context of illusions of judgment throughout chap. 3. Similarly, dealing with a risk that is subtle, hard to assess, complicated, a person will more readily fall back on a usual response—in more technical language, to a default response—that is, the response habitually triggered in hard-to-assess circumstances in ordinary life.

is seen as gain and what as loss. If accepting an increment to danger is seen as a loss from the status quo, the propensity to "better safe" will be very strong. But in the converse case, where it is giving up a bit of opportunity that is seen as the loss and obtaining a bit more safety as the gain, there is no longer any reason to expect the "better safe" perspective. So depending on how a situation is seen, it may be "waste not, want not" that is salient, or "better safe than sorry" that is salient. Hence Kahneman and Tversky's *framing* as well as my *fungibility* will shape a person's perceptions—or, allowing for the socially contagious character of such matters, they will often also shape a community's perceptions.

It may be that the framing is so unambiguously and forcefully cued that there is no practical room for further cognitive propensities to affect it. But for cases where both the danger and the costs of precautions (hence opportunities forgone by taking more precautions) are out of the range of ordinary experience, then what is framed as loss, what as gain, may be merely a by-product of how other things go. On the whole, "twoness" effects will more easily favor "better safe than sorry." It will be easier to trigger attention to a suggested new (or newly apprehended) danger than to trigger attention to equivalently subtle costs of precaution. Anchor-and-adjust effects (another major Kahneman and Tversky contribution) will have that effect when the risk involves some substance or practice that is firmly anchored in a sense of danger, as with radiation or asbestos or, as in more recent years, has become the case for smoking. On the other hand, if costs of averting danger as well as costs of accepting risk are on-screen, framing could go either way, or fail to be stable (as with the shifting perception of a gestalt drawing.)

4.12

Now on the account to this point, understanding expert/lay conflict requires no claim whatever about additional or different dimensions of risk that arouse the response of lay, but not expert, intuition. Rather, in terms of the risk matrix, what makes the expert visceral response to a risk different from the lay response is *not* some fundamental difference between how expert cognition works as compared to how human judgment in general works, but merely the tautological point that an expert by definition has a lot of experience on an issue where a lay person by definition is on unfamiliar ground. Every lay person on a matter at hand is an expert in various other things not at hand just now, and every expert on this matter is a lay participant with respect to most other matters.

But only some minority of issues become the focus of *stubborn* expert/lay conflicts. The more usual situation certainly is that lay opinion is content to leave things to people with experience. The extent to which that happens is largely unnoticed. If we do what seems comfortable, and it leads to no sharp challenge, then we do not explicitly notice what in fact we have done. The situation is just parallel to earlier comments on not noticing intuitions in general (3.2).

Think of actual risks we are all exposed to: in particular, think of actual risks that are sufficiently severe that we know we are talking about real lives at stake—not a risk merely conjectured or invisibly small, and certainly not a risk both conjectural and, even if real, invisibly small. Rather consider only activities in which everyone knows people actually are killed. Air travel and surgery are ready examples.

But here we have cases in which a social sense of fungibility is built into the situation. If activities are accepted where we know people are killed, then it can hardly escape being the case that "everyone knows" that there are nontrivial benefits associated with these activities. If not there would be controversy about permitting that to go on. Given that an activity that visibly kills people is permitted, it follows that there is apparently a social sense (an "everyone knows" sense) that there are substantial benefits to permitting that activity. So in the language I have been using, "everyone knows" that fungibility needs to be considered in considering restrictions of this activity.

But, as argued earlier (4.3), a sense of fungibility prompts looking for advice from people with special experience, which is to say, looking for expert advice. I have had the opportunity to test that quite a bit. Even in settings where everyone present is committed to the importance of public participation on issues of social risk, I practically never encounter favorable responses to a question about setting up a review panel of ordinary citizens for overseeing design of the air traffic control system or setting the procedures of hospital operating rooms. When things go wrong on such matters there is a demand for investigation and review, but that is most often done by experts, or else by elected or appointed representatives of the general public who focus on just those aspects of the issue on which there are substantial conflicts among experts. Deep distrust of experts is characteristically left to cases in which it is *not* clear that people are almost certain to be killed if poor choices are made.

4.13

Now put the points of the two preceding sections together. If a risk is subtle, it might be *framed* as a loss from the status quo, since the risks that are *not* likely to be bundled into the status quo are just those sufficiently subtle that if circumstances bring them to lay attention they could be apprehended as new risks. On the other hand, such risks could also always be seen as an aspect of some larger category of risk which is common knowledge. Since things could go either way, the framing will easily follow (hence reinforce) whatever prevails in terms of awareness of fungibility.

This yields the paradox that characteristically very small risks come to be the focus of salient lay concern as a loss from what had been taken to be the normal situation. Larger risks are known and either come to be taken for granted, or are also active matters of visceral concern among experts (hence not the focus of expert/lay conflicts.). But subtle risks can be brought to attention as new risks, not part of our usual sense of things. If fungibility has been missed, that will be the salient framing, reinforcing rather than conflicting with "better safe than sorry" intuitions. So although we have now noticed a second element that might account for expert/lay conflicts of intuition (framing as well as fungibility), framing turns out likely to be governed by fungibility: it would ordinarily contribute to the stubbornness of the conflict (should conflict arise) but would not easily be the essential basis of the conflict. For situations in which *fungibility* is missed, hence strong polarization easily takes hold, will also be situations in which *framing* will easily come to reinforce the polarized situation.

4.14

One further element remains to be considered. Fungibility and framing effects might be sufficient (as indeed I have been arguing) for an analysis of a person's intuitive sense that there is serious cause for worry even if experts claim otherwise. But those effects would not be sufficient for that person to give a plausibly good defense of why seeing the situation in that way makes sense. Unless there is something more to the story, a person would not expect to, or want to, justify his view of a matter by asserting that he is just ignoring the costs of precautions or that he is locked into a loss framing in a situation that logically could just as reasonably be seen in terms of a gains framing.

On a matter on which judgment is open to challenge, which is the situation by definition in a matter of public controversy, a person ordinarily feels uncomfortable unless he can point to reasoning—why that justifies (for missed fungibility) a firm lack of interest in considering a trade-off between the costs of accepting the risk and the costs of precautions. And along with that (for a loss framing) a person feels uncomfortable with nothing to point to as a good reason for seeing some very small risk (on the scale of risks routinely treated as negligible) as too important to ignore. Consequently, a person will find that stubborn adherence to his intuition is embarrassing without some further support: and an activist on an issue (hoping to persuade others to share his view) will especially want to avoid that.

Consequently, we want to consider whether there is something about conspicuously stubborn situations that helps defend intuitions logically vulnerable to attack. We could expect, for conflicts of expert/lay intuition that prove to be particularly stubborn, that it will somehow be especially easy to see things in a way that can *justify* treating some kinds of costs on a different scale from others. The "inversion" exercise that takes up the bulk of chapter 5 is ultimately concerned with ferreting out how that defense might operate.

Experts and Victims

5

This chapter provides an exercise in what I will call "inverting" Covello's list of psychometric dimensions (fig. 2.1, summarized here).* The usual account explains expert/lay conflict by lay sensitivity to extra dimensions of risk—such as those in Covello's list—combined with expert insensitivity to anything beyond expected fatalities. I will pursue the opposite view here. I argued in chapter 2 (esp. 2.8) that the extra dimensions could at least as plausibly reflect rather than cause lay concern. And we now have in hand an account, or at least a conjecture, that could explain the observation of stubborn expert/lay conflicts without giving these extra psychometric dimensions a fundamental role.

By the end of the chapter I will have said something more detailed about the subordinate role that would then be left for the psychometric dimensions. But the main task of this chapter is to see how far we can in fact explain the psychometric dimensions as a *by-product* of visceral responses to the risk matrix considerations. That is what I mean by *inverting* the psychometric dimensions.

For example, given the ubiquity of "major or minor" accidents,

*From fig. 2.1: (1) catastrophic potential; (2) familiarity; (3) understanding; (4) uncertainty; (5) controllability; (6) voluntariness of exposure; (7) effects on children; (8) effects manifestation; (9) effects on future generations; (10) victim identity; (11) dread; (12) trust in institutions; (13) media attention; (14) accident history; (15) equity; (16) benefits; (17) reversibility; (18) personal stake; (19) origin (human vs. natural).

item 14 is about the salience of accidents associated with some risk, which (as mentioned in 2.8) is not necessarily closely tied to how often or how seriously accidents actually occur. A strong visceral response to some risk will prime a person to fix in memory impressions of accidents linked to that risk. And if that visceral response were widely shared, that would prompt close press attention to even minor accidents, which would make more accident impressions available that might contribute to memory. The cumulative effect would be that, if some risk typically prompts visceral concern (for any reason), a by-product of that concern is likely to be a sense that activity associated with that risk is accident-prone. And that would remain likely even if the actual accident record looks very good compared to other activities not subjected to such scrutiny.

Of course, if an activity actually has a bad accident record, all this could be dismissed (and obviously should be dismissed) as secondary to the point that an activity with a bad accident record deserves to be treated with unusual concern that things may go wrong. So an elevated sense of accidents in connection with some activity could be caused by a strong visceral sense of risk, or it could cause that visceral sense of risk, or something in between. But in the psychometric reports, accidents are commonly associated with activities like the transport of nuclear waste, which, on the record, have exceptionally favorable accident histories. So the possibility of a confusion between cause and effect is hardly a remote one here. Sometimes, at least, it certainly occurs. As already partly sketched (2.8–2.10), less obvious parallel arguments arise for every one of the 19 items on Covello's list.

But whatever the connection between the psychometric dimensions and visceral concern, there must still be a connection (perhaps indirect) that somehow makes sense in terms of what makes a risk seem especially worrisome; or else how could we get the persistent statistical associations? Some connection must hold whether or not (or to whatever extent) causality goes in the reverse direction from the usual interpretation of the psychometric work. On the usual interpretation, the psychometric studies are discovering what characteristics of a risk arouse concern. The inverted effort here does the opposite. We want to see how far we can turn Covello's list inside out, making the psychometric dimensions contingent on just the points that emerged as important on the risk matrix argument of chapter 4.

Accordingly, we will be looking for explanations of the psychometric dimensions tied to the three categories that emerged in chapter 4: reflecting fungibility (or its absence), reflecting framing (of the status quo), and supporting reasoning-why that might defend one-sided "better safe than sorry" responses even if evidence goes mainly the other way.

5.2

As background, we want to have at hand some further results from the psychological literature. For example, absent close feedback, intuition seems usually governed by some small set of cues.[1] This is subjectively surprising, since our introspective sense of how we choose is usually quite different. Yet several decades of studies have shown this tendency to be remarkably strong even for experts (see Dawes 1979; Dawes and Corrigan 1974). The work has mainly compared the judgments of medical and psychological clinicians to simple linear models. But less detailed work on other sorts of expertise has yielded similar results. For inexperienced judgment, the modestly few dimensions easily collapses to a *very* small effective set of dimensions. We fall back to widely available general cues, sufficient to prompt some response, perhaps only a default response.

But if we ask for a person's subjective sense about what governs judgment, we get the reverse of this result. We see ourselves making judgments that reflect refined consideration of interactions among a rich array of dimensions. And indeed, in contexts where we have a lot of experience, and with good feedback to alert us to how well we are doing, that seems to be accurate: then the person making the judgment can only partially and crudely explicate the subtlety of her own responses.

In such rich-feedback situations it will be hard to build a computer model of expert judgment (e.g., for chess, despite the artificially simple world of a board game).[2] But in situations without

1. On "implicit learning," see esp. Reber (1992), as well as Dawes (1979).
2. The logic of chess is simple, but the combinatorial explosion as sequences of future moves are considered is overwhelming. Strong computer chess programs rely primarily on the machine's capacity for brute force explorations

prompt reliable feedback, building an effective model has turned out to be unexpectedly easy, as reviewed in the Dawes papers. Judgments can be matched to the use of a very few cues (even though the actor thinks he is being far more subtle), and the computer can do better by actually using, if only in a very crude way, the full range of cues the chooser thinks of himself as using, but does not.

Even after several decades, these results still seem surprising. But, however surprising, the basic result has been replicated many times and in a wide range of situations. These results reinforce the reasons already discussed (2.8) for doubting the rival rationalities accounts of expert/lay conflicts. It is implausible on its face to suppose that lay judgment with respect to risks far outside the realm of normal experience could actually be sensitive to anything like the 19 items that show up as significant in the psychometric work. But the constructive side of that, as suggested already, is that by "inverting" Covello's list (looking at responses as possibly *prompted by* rather than *prompting* perceived risk) we can hope to make sense of the psychometric results in terms of the much more parsimonious risk matrix argument.

Reviewing: we think of ourselves making judgments in refined and discriminating ways. But at least in contexts that approximate the conditions of concern here (where there is weak or nonexistent feedback from judgment to consequences) it ordinarily turns out that a far simpler account is adequate to explain the great majority of judgments. And we now in fact have in hand the means to test a conjecture about how a simpler account might work.

5.3

Inverting the psychometric dimensions turns next on a cognitive point, first mentioned in 2.7, which itself derives from a more

---

of superhumanly vast numbers of combinatorial possibilities. But computers can rarely (as of this writing anyway) defeat the best human players, who rely on tacit intuitive insight into the patterns they somehow apprehend but can only crudely describe. That tacit intuition is substantially contingent on extensive prior study and analysis, but at the moment of choice what makes the difference between the world class player and an ordinary player is a sense of things he can feel but not articulate.

general argument spelled out by Nisbett and Wilson (1977). If we typically perceive some activity or substance as risky (i.e., we respond to it with a marked visceral sense of risk), then, if asked to rank it on various dimensions, we will report some set of responses that fit with that sense of visceral risk. Some (conceivably each and every) element of such a report might in fact coincide with the actual influence of those dimensions on the individual's visceral response. But absent confirming evidence, that can only be a conjecture. For where it has been possible to apply careful control—which usually means in constructed situations, as reviewed by Nisbett and Wilson—it turns out that subjective reports about what governed judgment commonly bear no relation to what we know must have accounted for the judgments.

Women asked to select the best hose from a choice among four actually identical pairs usually chose whatever pair was put on the right-hand side of the array. Since there was in fact nothing actually to choose among (the hose being identical), it would not be surprising that some trivial advantage—here an advantage to being rightmost, from our left-to-right reading habits—would bias choices that way. Stage conjurors routinely take advantage of such biases in how people choose. But the choosers did not report the choice as arbitrary. (So why not choose the right-hand pair?) Rather, they reported noticing qualitative differences of weave, color, and so, on which might have been very reasonable reasons for choice, except that in this controlled situation we know that there were no such differences.

In numerous other experiments, parallel results were obtained from men as well as women, workers as well as students, and (in the work discussed) for expert as well as lay judgments.[3] When asked about the effect of the cues that, from the artificial setup, we know actually controlled the judgment, the response is very often the subjects' confident denial that those cues could have had any such influence. Critiques of Nisbett and Wilson have raised

3. But if expert as well as lay judgment is subject to all the odd features stressed here, why suppose that expert judgment is any better than lay? Because on a matter where a person has a lot of experience, the anchor for intuition will be far more precise than for a person without experience, though the adjustments will be subject to the same (qualitative) kinds of variance.

doubts on various points of detail, but nothing that would affect the argument here. Their bottom line remains strong: what we report as causes are what from experience we sense as plausible associations with the judgments we make—which may or may not have anything to do with what was actually governing in the case at hand. And in fact unless the judgments are based on extensive experience with close feedback, our reports will almost certainly be inaccurate, and often grossly so.

Consider, then, a case in which I have a strong visceral sense of risk with respect to some matter that lies well outside the range of direct observation (so I necessarily will lack feedback from judgment to consequences). Now I am asked to rate that matter on a list of dimensions (see the note above that reprints Covello's list from fig. 2.1). Translating the Nisbett and Wilson argument to this context, we can expect that I will see the dimensions offered in some way that fits my overall response, but not reliably in a way that reveals what, in fact, governed my response.[4]

The resulting fit can be thought of as akin to an investment portfolio. Two people with the same intensity of visceral risk might choose quite different "portfolios" in a study that offers a large array of possible dimensions, as two individuals with equal resources and also equal risk aversion might choose quite different portfolios of investments. With many options there will be many ways to formulate a roughly equivalent package.[5]

So if subjects are offered only two dimensions (say, dread and

4. The risk of cause/effect reversals like those demonstrated by the Nisbett and Wilson experiments is commonly ignored even when that is easily likely to be relevant. So discussions of postelection polling commonly take it for granted that what voters report as their reasons for choice were what in fact guided their choices. But a person who voted for a Republican who talked a great deal about inflation is likely to report inflation as his special concern whether it was in fact determining or not. Not many people would report that they voted Republican this time because they always vote Republican. But political managers have learned to distinguish between what gives people something to explain their vote and what in fact drives their choice. Hence the important role of focus groups, where you can see where people's passions are engaged, as against where respondents are only repeating what seems to make sense for a person voting as they are likely to vote.

5. I owe the portfolio analogy to Don Coursey. A consequence is that responses are spread widely. Hence the correlation with perceived risk of any particular item on a list like that of fig. 2.1 is typically weak. And the pattern of responses varies across items from one study to another: for across studies there

familiarity) both would show up as highly significant. But if offered anything like the full array of figure 5.1, each may remain statistically significant (given a large sample of respondents), but now with only mild (imputed) effects on the sense of risk. With more items to choose from, we pick a broader portfolio, "investing" less in any particular item. But all the dimensions called into play might be tied to some small set of controlling elements, which is the possibility we want to consider here. On the risk matrix view, we expect to find an overriding importance of fungibility, and a secondary but still strong role of framing. And we can expect to find some additional factors tied to defending the visceral response as a reasonable response.

In the actual psychometric work, subjects are not directly asked to give reasons why they are worried about some risks and not about others. Rather as described in 2.8, subjects are asked to rate various risks overall and, ordinarily separately, they are asked to rate the risks along various dimensions. The psychometric results correlate intensity of perceived risk against the intensity of response along the various dimensions. Subjects, consequently, do not themselves make explicit assertions about what they think is governing their perception of risk. But the dimensions we get turn out to be ones that look intuitively reasonable to the researchers and their audiences. Yet, as Nisbett and Wilson stress by reporting their own unreliability in foreseeing what cues would in fact govern their subjects' responses, the actual connection between cues and responses is just as opaque to researchers as it is to subjects. So that the correlations look plausible does not demonstrate that they are causally correct.

Summing up: when we turn from questions of what people reveal as subjective correlates of strong perceived risk to what actually governs such perceptions, the 19 significant dimensions listed in figure 2.1 look, on the face of things, far too large.[6] On

---

are differences in characteristics of people surveyed, in the array of risks, in the presentation of those risks, and in the number, and, to some extent, the type of dimensions offered. The most detailed treatment of the psychometric dimensions I have encountered dates from the work of Vlek and Stallen (1980), which appears quite early in the development of these ideas.

6. I cannot claim that such a list of causal factors is impossibly large: it could be that some plausibly small set of factors govern any particular case, but

the other hand, from the risk matrix analysis (chap. 4), an account of what is actually governing visceral responses, may indeed be very much simpler. Two direct elements were developed in chapter 4: an absence of perceived *fungibility* and a loss *framing* of the risk. But if the former held, we found good reason to expect that the latter would fall into place as support. And at the end of the chapter we also took note of perceived reasonableness. All will play large roles here, but with fungibility at the heart of things.

5.4

If we start from the discussion in 2.8 of Covello's list,[7] the most obvious item is benefits (item 16). If there were no benefits from accepting a risk, then by definition a person could only be in the "better safe than sorry" cell 2—or, lacking also any perception of nontrivial losses that go with the risk, in the off-screen cell 4. As usual in this discussion, we would want to consider to what extent benefits are in fact small (and whether they are small relative to whatever risk is in the situation), or whether a low sense of benefits more plausibly responds to perceived riskiness than accounts for it. The benefits of nuclear electricity are sensed as smaller than the benefits of electricity from coal. If this is because the risks of nuclear are felt to be greater than the risks of coal (And what else could account for such a difference?), then exactly what we are seeing is a case where perceived risk prompts the psychometric response. Indeed, on this particular dimension, inverse causation is not even controversial, since the effect has been documented within the psychometric work (see Alhakami and Slovic 1994).

Much else in the list of dimensions relates to fungibility (or

---

across a long list of risks and many respondents, this accumulates to a large number of dimensions that play a substantial causal role somewhere. But that is not what Dawes's discussion of simple linear models suggests. So while 19 statistically significant dimensions taken to be causal is not logically impossible, it does seem to me highly improbable.

7. As is explicit in the right column of fig. 2.1, for many of the dimensions it is the absence of the quality that aggravates visceral risk.

its absence): some items subtly so, but others in ways almost as direct as benefits. Of particular importance, as has often been noticed, is voluntariness. If a risk is felt as voluntary (item 6), then on the fungibility argument it is tautological that we would not find that lay intuition sees that risk as worrisome but experts do not. Instead, the expert/lay conflicts to expect will be of the opposite sort, with lay concern low about a risk that the experts think is substantial. Commonplace examples would involve smoking, seat belts, or safety helmets. Where we can notice contrary situations, they are characteristically not stubborn conflicts, but transient conflicts, like concern over the emissions from microwave ovens or cellular telephones.[8]

For if exposure to a risk is seen as voluntary, then apparently there are perceptible benefits (or why would you be doing whatever it is that entails the risk?) and apparently the costs that go with that risk seem commensurate with the benefits. If there was some way of reducing the risk that you judge worth its cost (in

---

8. Microwave ovens in fact provide two examples. The initial concern was about the ovens in home use. Consistent with the argument here, since this would be a clearly voluntary choice, fungibility soon asserted itself. A social consensus then emerged that, with quite trivial reassurances about industry standards, the risk was negligible. But that did not entirely quiet concern. Demands emerged that the ovens be barred from public places, where a person might be involuntarily exposed. However, since the social view was that concern about microwave ovens was excessive, many more people saw inconvenience from such a ban than saw reassurance. The result was pro forma regulation that allowed officials and legislators to respond to concerns without imposing any substantial costs on anyone.

Mandated warnings can still be found in places like gas station minimarkets, where microwave ovens are available for customers to warm up snacks. The presence of these signs reflects—indeed twice over—the transient nature of the microwave concern: the warning notice reflects failure of initial efforts to ban use in public areas, and their effective invisibility (when asked inside a store, a large fraction of people are unaware of the notice though it was plainly in sight when they came in the door) reflects the passing of interest in the whole matter. It becomes hard to find anyone who can remember being much concerned. For by now "everyone knows" that it is not worth worrying about.

A similar, and in various ways even more striking tale can be told for proposals to ban saccharin: in this case government action consistent with what the public is perceived to demand for other food additives was overridden by public demand for this (by the usual regulatory criteria) carcinogenic food additive. The issue generated an enormous volume of mail to Congressional offices, overwhelmingly against the FDA's proposed ban, as will be reviewed in chap. 6.

dollars, or inconvenience, or lost excitement, etc.), why have you not adopted it? So, just to the extent that an activity is seen as voluntary, it takes unusual circumstances for a person to feel blocked in the cautious "better safe than sorry" state and unable to move to a more comfortable choice.

Of course there are cases of on-alert (high visceral sense of risk) voluntary activity, like rock climbing. Then a person is sensitized— ready to back off to "better safe" if necessary. But this concerns tactical situations a person knows he will sometimes encounter in an activity which, in a strategic sense, he regards as making sense. His strategic sense of participation is not "better safe" for the activity overall or he would not be doing it voluntarily.

In general, fungibility comes in as soon as I see myself as having a choice about how far to compromise some activity that I value. Further, if something new about the situation prompts a sense of abnormal risk (the precondition for the visceral sense of risk [2.9]), I cannot escape some awareness that a trade-off is needed to choose what will make sense for me. In such circumstances it would be odd that some merely ephemeral or weakly conjectured or extremely remote risk could prompt a person to a more than transient sense of danger. If (somehow) a persistent concern did take hold, the person is likely to judge his own response to be unreasonable and to seek help from a friend or a therapist. So actual voluntariness relates in an immediate and transparent way to the driving feature (fungibility or its absence) for interpreting responses in terms of the risk matrix.

But as with benefits and with items on this list generally, we can expect a tendency, given a perception of danger but no parallel perception of opportunities that might have to be given up to avoid the risk, a person will have an enhanced propensity to see the activity as involuntary; and that propensity disappears if costs of precaution as well as danger are on-screen. In the first case, fungibility is absent; in the second we do have fungibility. If I feel comfortable flying, I am also likely to feel that flying is voluntary; if I have become worried about air traffic safety, flying starts to feel involuntary. I become alert to how I really have no choice but to fly if I want that vacation in the Caribbean.

5.5

Framing, the second of the two key features discussed in chapter 4, comes as immediately into play as fungibility, most obviously through familiarity (item 2). "Familiar" is commonly used as a synonym for "customary, accepted, routine" (all terms so listed in my thesaurus). So a risk perceived as familiar will, by definition, be framed as part of the status quo. Costs that go with a risk-reduction proposal connected with a familiar risk will consequently be seen as costs incurred (losses accepted) in the hope of making a gain (explicitly, to make things safer than I am accustomed to).[9] If there is expert/lay conflict, lay intuition will then more commonly resist rather than run ahead of expert concern. So it took years of effort to persuade the lay public to take seriously the experts' urging to wear seat belts, stop smoking, and so on. But the same sort of costs proposed to defend against some new—or merely newly recognized—risk (some risk not part of my familiar sense of what goes on here) will be costs incurred to avoid a loss, and then loss/gain asymmetry would bias judgment toward "better safe than sorry."

Yet a perceived loss framing can be the *effect* rather than the *cause* of a "better safe" sense of a situation. An activity or situation might be familiar, but many details (a particular sort of risk, or aspect or risk, or the mechanism through which that particular sort or aspect of risk works) will be unfamiliar. So, depending on the focus of attention, a risk might seem either familiar (e.g., as part of a familiar activity, like trucking) or not (as an aspect of trucking risk that I have not thought about before, which is rich field since there are few particular aspects of trucking I will have thought about).

Hence, framing and fungibility must be easily entangled. If

9. Again (recalling the rock-climber comment a moment ago) we have to allow for sensitized cases, where a person knows that an activity is dangerous by his own normal standards but finds it sufficiently satisfying that he wants to do it anyway. So the activity can be familiar as well as voluntary, but a chance to diminish the risks will be more immediately attractive than would be the case for more commonplace activities. "Satisfaction" is not necessarily defined in a narrowly hedonistic sense: consider the resistance fighter.

fungibility is missing because a person sees opportunity as negligible compared to danger, that will almost automatically be accompanied by a framing that sees that danger as a loss from the way things are now. And the converse is true if a person sees opportunity but no significant danger. Any danger can always be seen as part of a broader class of hazards that have always been around; and any activity, however well established, can be seen in terms of some newly noticed feature (such as secondhand smoke in the case of cigarette hazards) that makes it feel like a disturbance of the status quo. You can see pesticides as a new risk (man-made pesticides are indeed new), or as a minor part of what has always been around (every plant is equipped with many natural pesticides). Hence framing as an *independent* element will be important in contexts of what we might call a *normal* controversy—a controversy not marked by sharply polarized views. But in polarized controversies, framing is more likely to be a by-product of risk matrix effects than an independent impression. But, of course, once in place that would become a substantial reinforcer of the polarization, since actors on either side will be prone to contrary framing intuitions.

Even aside from that argument, voluntariness (acting as a proxy for fungibility) and familiarity easily come to be intertwined. If something is part of my usual routine and hence familiar, it will also come to seem voluntary, since it ordinarily does not even occur to me to do anything different. A person may come to see stopping at a red light as voluntary, even when there is no other car in sight. If challenged, this can be rationalized as prudent behavior that the driver voluntarily chooses, or at least as a prudent habit that is not worth rethinking even if there is no car in sight. In terms of survey responses, the interaction between voluntariness and familiarity might be diluted by the "portfolio" effects mentioned earlier. An extreme provoker of visceral risk (like nuclear waste) might prompt extreme ratings on both dimensions (very involuntary and very unfamiliar), but with less extreme perceptions of risk a high rating on voluntariness would tend to crowd out familiarity and vice versa.

5.6

Various other items on Covello's list fall into place as secondary features linked to fungibility or familiarity or to both. We come to treat things we familiarly use as things we understand, so that vitamin pills, VCRs, and so on are not likely to rate low on understanding (item 3), even though few people could give even a vaguely correct account of how they work. On the other hand, an unfamiliar thing will easily rate low on understanding, even if a careful effort has been made to explain how it works,[10] so that something (or an aspect of something) that prompts a "better safe" sense will easily be seen as low on understanding. Controllability (item 5) is really a subcase of voluntariness. If the risk is controllable, then at least some aspect of the risk (namely, whatever it is I see myself as controlling) is voluntary. Alternatively, "controllability" usually entails familiarity. A person just learning to ski or bicycle is in one sense in control (no one else is, surely) but, in a stronger sense, does not feel in control.

Benefits (item 16) too subtle to be easily seen are likely to be taken as obvious if social knowledge prompts confidence that it is there. For familiar activities (like taking a vitamin pill), a person would feel that "everyone knows" there are benefits in this. Contrariwise, for a risk that goes with an unfamiliar activity, benefits are seen as speculative if they are seen at all. And origins (item 19) links to familiarity in a tangential but nevertheless important way: even if I am not personally familiar with the risk, it is familiar in a larger sense of being something that (presumptively) has always been around, is the usual state of things, and so on. And familiarity ties to framing the way already discussed.

So in one way or another, six of the 19 dimensions immediately fit into the risk matrix story as items that easily relate to

10. Many people (before one-step programming became commonly available) would say they did not understand how a VCR works, but meaning only that they did not know how to carry out the mechanical steps to program the device, not that they felt they did not understand what happened once the buttons were pushed. For even though the technical aspects may have been completely beyond their comprehension, that the machine somehow worked was too familiar to be puzzling.

fungibility and framing in the expected way. If danger is on-screen but opportunity not so, then this absence of fungibility will encourage perceptions along these dimensions that rationalize and (by influencing framing) also reinforce "better safe" intuitions.

Media attention (item 13), on the other hand, is not a reason-why at all, but a shaper of attention and of the sense of what "everyone knows," but also a reflector of what "everyone knows." So in part press attention will be an independent factor, stimulating concern (or quieting it) about some risk, not merely reflecting widespread visceral concern. Yet by far the largest single factor in what the press attends to (since success is so heavily contingent on audience size) is what the public wants to hear about. But if issues get a person's attention only through social processes (to which the press is both responsive and a fundamental contributor) then a person would naturally have some sense that things he worries about are usually things the press pays attention to. If a matter has become a matter of controversy due to widespread public concern, it is tautological that there will be press attention to it. So media attention introduces another category: a psychometric dimension that does not of itself rationalize strong "better safe" intuitions, but nevertheless fits readily into the risk matrix account since it is an essentially inevitable correlate of expert/lay controversy.

So on this first sweep we noticed six items that in some obvious way link to fungibility or framing or both (benefits, accidents, voluntariness, familiarity, controllability, and understanding) plus one other item (media attention) that to some extent is a cause and to some extent an effect of expert/lay conflict, but in any case will be automatically present in the case of stubborn controversy.

Twelve items remain. In part we can expect these to relate in some less immediate way to fungibility or framing. But as suggested at the close of chapter 4, we should also find features which link to quite another aspect of the situation. They should somehow provide a package that will serve to defend lay intuition: items which in one way or another have the property that they would help a person feel their visceral sense of the situation is *right*, despite contrary expert assessment.

5.7

Dread (item 11) in one sense has a distinct qualitative character as mentioned in 2.8; this is the sense in which dread conveys (from my thesaurus) "fright, terror, horror"; but it also can be simply a relabeling of the visceral sense of risk the list is purporting to explain. My thesaurus lists, as a verb, "distrust, fear, suspect," or, as a noun, "anxiety, apprehension, trepidation, uneasiness."

If I have a strong sense of dread in the sense of horror, such as occurs when people dread snakes, heights, and so on, then that will of course directly affect my perception of risk. But the extended senses of dread (as anxiety, apprehension, etc.) makes more sense in the context of actual cases. For example carcinogens are treated with indifference in some contexts where fungibility is present, as with exposure to radiation from a dental X ray or from visiting a resort at high altitude (such as Santa Fe). In contrast, a person who dreads snakes or flying does not exhibit indifference if circumstances make exposure unavoidable. So an elevated sense of dread for the same risk lacking fungibility appears to reflect the more general sense, where dread is synonymous with the visceral response it is called out to explain.

Whether dread is a substantial component of concern about, say, pesticide residues will depend partly on what context is tacitly imputed. A pesticide residue from a commercial grower— especially a commercial grower in a remote area, so there will be no local press attention to the plight of the grower—may be dreaded, but perhaps not the pesticide from your children's garden. And the sense of dread will also vary with what other dimensions of risk are offered (recall the portfolio argument discussed in 5.3).

In any case, since dread is often a synonym for visceral concern, it correlates, of course, with visceral concern. Dread is something that a person can report anytime he senses more visceral concern than he can otherwise explain. And even more clearly this also holds for (dis)trust (item 12). But that has already discussed at some length (2.6).

Accidents (item 14) and the relation to ready media attention also have already been discussed (5.1), so that even otherwise minor accidents are seen as newsworthy. But seeing a matter as accident-prone also rationalizes doubts that the expert can foresee what might occur.

Several other items seem harder to see as *causes* of visceral concern, but readily come into play as elements that can always be appealed to *rationalize* concern, even in a case where no evidence of substantial actual risk exists.[11] At least some subset of these items is always available to fill out a person's portfolio. They include (for varying reasons already mentioned [2.8]) uncertainty (item 4), catastrophic potential (item 1), delayed effects (item 8), and effects on future generations (item 9). All were discussed in chapter 2 as items that could rationalize concern even if no harm could be detected or reasonably inferred.

Effects on children (item 7) and personal stake (item 18) involve different but not incoherent uses as rationalizers. No explicit explanation at all is required to justify special caution if a person sees himself as directly at risk, but if he does not there are ample options left in the list (items 1, 8, 9, and 15, at least) to express empathy for cases that seem risky but do not involve any personal stake. Further, a person might honestly and indeed reasonably feel some personal stake in a matter that does not effect her in a merely selfish way. Children (naturally) are seen as especially vulnerable and especially warranting protection.

Finally, rounding out this group, a perceived absence of reversibility (item 17) could only enhance any propensity toward "better safe than sorry," though in some sense all choices are irreversible, since we can never turn back the clock. But we don't often think of breaking open a peanut as irreversible.

5.8

Now consider an item that plays a special role: equity (item 15), or, more exactly, since (parallel to voluntariness and some other

11. Perhaps this should say "reasonable" evidence, to take account of the fact that there is no position for which no evidence at all can be offered. There are people—in fact, clearly sane people, sincere in their beliefs, however stupid or disgusting their views may seem to the rest of us—who are prepared to go on at great length on the evidence for doubting the Earth is round, or that the Holocaust ever occurred. They have what they see as ample good evidence, but

items) it is concern about the absence of equity that plays a role here. Directly and indirectly, many of the other items link to it.

Children (item 5) form the broadest class of people conspicuously vulnerable to unfair treatment, including both sexes and all races. In particular a person might be especially concerned about risks to children, who are naturally envisioned as especially vulnerable and also, since they are just starting life, as having a lot to lose. And future generations (item 9) must depend on us for fair treatment, since there is no one else to care about their interests. But parallel to familiarity and voluntariness, equity also links to various other items whose presence as significant contributors to perceived risk might otherwise be puzzling.

Suppose that for one reason or another harmful effects will occur in some presently unforeseeable context. That would raise equity problems. A number of items fit with that. Perhaps the risks are not fully understood (item 3), or they are delayed (item 8 or item 9), or uncertain (item 4), so they might prove more dangerous than anticipated and damage victims not recognized to be at risk. Or perhaps the risk is of some catastrophe (item 1) that produces great harm but in a concentrated time and place no one can now foresee. In any of these case those who might suffer harm obviously have not been given fair warning, since no one knows where the danger may strike. If the danger is involuntary (item 6), then prima facie, those exposed to the risk have not accepted it, and an absence of clear benefits to those exposed, suggests, on its face, inequity.

Identifiable victims (item 10)—this completes Covello's list of 19 dimensions—will easily be suffering more harm than they deserve (we all run the risk, but a few people get the brunt of the actual damage). And if the origin is not natural (item 19), then there is someone who might be blamed, a usual if not strictly necessarily condition for an injustice. And related to that, is the dimension of accidents (item 14). Since there is usually someone to blame when an accident occurs, a perception of accident-proneness fits nicely with a perception that this is an activity in which culpable actors harm innocent victims. Overall, equity (or

---

which the rest of us are likely to see as trivial, out of context, unsupported, fraudulent, etc.

lack of it) provides a common thread that links many of the usual risk dimensions, including several items (like delayed harm as opposed to immediate harm) for which the linkage to visceral risk might otherwise seem thin.

But equity has a different character from any of the other items, with an only partial qualification for dread (5.8). The crucial negative side of this term (inequity) cannot be value neutral, and also makes no sense except as a term about visceral response in some dimensions *beyond* the tautological "better safe" feeling of the situation. Even the other items that readily *link* to inequity can be understood (artificially for some, but still, in various contexts, it might be that way) as mere descriptions of states of the world without a valuative component.

Consider even risk to children, which easily links to equity but need not always evoke it. There are risks to children in vaccinations. Some will get sick from the vaccine, and nearly all will find getting the vaccine an unpleasant experience—for some it will be a terrifying experience. Nevertheless it would seem odd to most of us to think of children subjected to vaccination with the feeling of moral resentment that is built into the notion of inequity. Inequity is not something that sometimes, or even often, carries a negative emotional color. "Inequity" as much as, for example, "disgusting" or "despicable" cannot exist except with a negative emotional color.

But once a situation is perceived as unjust, it is difficult— indeed it prompts a further sense of revulsion—to subject it to cold calculation of costs against benefits or of comparative risk. But such calculations are just what experts have to offer. So what gets reported as a lack of trust in authorities can reflect (or reflect in addition) something very different: an inability of authorities whose claim is that they are relying on the best expert advice to respond to at least one ingredient reinforcing a "better safe than sorry" intuition. It is always hard for mere logic to overcome sharp intuition, as most readers will have experienced firsthand in connection with the chips puzzle of chapter 3. The compounding effect of a strong emotional response is then a substantial matter. Resistance moves from being hard to overcome, to being a matter of principle that a person might only with shame *allow* to be overcome. And all the more easily so if the costs of taking more care are hard to see, or whoever pays can be seen as *guilty* and deserves to pay independent of any detailed claims about costs and benefits. Indeed, whatever mere logic might recommend, the substance itself becomes not only dangerous, but guilty, contaminating anything that it touches.

The net effect, in the language used to title this chapter, is that experts appear to be aligned not just against lay intuition but against *victims*. A noted psychologist argues:

> Hot cognitions . . . seldom [seem] subjectively false. Once formed an [affective] evaluation is not readily revoked. Experiments on the perseverance effect, the strong primacy effects in impression formation, and the fact that attitudes are virtually impervious to persuasion by communication all attest to the robust strength and permanence of affect. Affect often persists after a complete invalidation of its original cognitive basis, as in the case of the perseverance phenomenon when a subject is told that an initial experience of success or failure has been totally fabricated by the experimenter. (Zajonc 1980, 157)

We could qualify Zajonc's argument a good deal without seriously compromising the point relevant here: that affect ("hot" cognition, emotion) tends to lock in a response. In particular, Zajonc's argument emphasizes the likely significance of the "victims" dimension, which both helps lock in the visceral response and helps lock out a sense that what the experts have to be say needs to be listened to.

5.9

It is clear by now that we can give an account of the psychometric dimensions in terms of what we can call the three Fs: fungibility, framing, and fairness. But I have allowed (4.4) that we can also construct a "usual story" account of the three Fs. The three Fs on that view would become a way to interpret the psychometric story, not an alternative to it.

But it is already clear that I want to argue for a strong interpretation of the three Fs that turns the usual view upside down. On this account, two of the three Fs—fairness and framing—themselves would ordinarily be derivative. Fairness and framing would remain important, but usually only because they help lock in conflicting intuitions that are created when experts see both danger and costs, but most people who are not expert sense danger but miss costs. In such cases, experts start from the fungible cell 1 of the risk matrix, but lay intuition is in "better safe than sorry" cell 2. The predominant expert intuition then will usually be driven (by the dynamics described in 4.3) to "waste not, want not" (cell 3), and we reach the sharply polarized expert/lay conflict of intuition that concerns us.

In just such cases, framing and fairness intuitions are likely to be highly vulnerable to merely conforming to (and so also reinforcing, but not governing) what makes a person more comfortable with a "better safe" sense of the matter. That ultimately turns on the point that expert/lay conflicts in which experts think lay judgment is overreacting to risks are almost bound to be cases of *subtle* risks (4.11). If the risk is not subtle, but clear—so we can *see* people are being killed, not just worry or suspect that is happening—then almost always that will also be an activity that "everyone knows" cannot be restrained except at a real cost. Then "everyone knows" that both danger and costs of meliorating the danger exist, so by definition everyone starts from cell 1. There is no absence of fungibility for either expert or lay judges. On the other hand, if there is social controversy about a risk that is not subtle, then it is not at all likely that there would be a strong expert consensus that sufficient precautions are being taken. Experts in epidemiology, or statistics, or any other relevant field, may converge on a consensus view of the statistical risk, but in this kind of case (where the risk is not subtle) those experts will usually diverge on the social and ethical intuitions that lie behind visceral judgments of how safe is safe enough. So we might have intense controversy, but not controversy sharply polarized between expert and lay views. Sharply polarized expert/lay conflicts, to repeat that important point, are characteristically about subtle risks.

Reviewing the argument sketched at the end of chapter 4: a situation that implies a subtle risk will (by definition) lack sharp, readily perceived, routinely processed information. Were that information present, it would likely make one way of framing the status quo dominant; the same, perhaps, could be true for intuitions about what would be fair. But since the situation is subtle and not clear, people worried about a risk will be primed for fairness and framing intuitions that help make sense of and help defend the danger intuition. That will be so whatever the reason for the danger intuition—perhaps a perfectly good reason, but equally so if the reason happens to instead be perfectly bad.

As noticed at the conclusion of chapter 4, a subtle risk can always be seen as a loss from the status quo even if it has always been around, since it is not something a person has ever been able to notice. So framing easily falls in line with and reinforces the absence of fungibility. Then any possibility, and inevitably there will be some, that can rationalize the concern that the experts might be wrong will loom large and dark. And given the imperfection of human beings and human institutions, it will also ordinarily be the case that points can be noticed where people have been, or at least might have been, victimized by careless or ruthless or greedy behavior. Then nothing more may be needed to make counterarguments about costs and benefits seem not merely irrelevant, but perverse.

5.10

If the usual story is right, the public is getting the sort of environmental regulation it wants and also the sort that democratic principles would prescribe as just. Whether experts see things that way or not, the public (if this view were right) cares about the psychometric dimensions, and in fact cares so much about them that a democratic government need have little concern about mere efficiency in the way it regulates risk. On the other hand, if our intuitions about risk are often like our intuitions in response to the chips puzzle—clear and emphatic but nevertheless often illusory and bound sooner or later to lead to disillusionment if acted on—then it not only misuses resources but it ultimately undermines democratic institutions for government to do business in a way that eventually will come to look perverse.

Examining Cases

6

6

Somehow—exactly how lies at the frontier of work in social psychology, sociology, and political science—it comes about that certain ideas take hold as things "everyone knows." Or more exactly, "everyone" in the reference group for a particular individual. Defining what becomes a person's reference group (which can even vary for the same person contingent on the issue) is itself a challenge. But roughly, it is how "people like me" see things. By and large, however, what "everyone knows" is what it would be advantageous for that individual to believe. And yet some of the things we come to know are open to grave doubt (even some of the things that are nevertheless advantageous to the individual to believe). So what "everyone knows" sometimes eventually comes to be what "everyone knows" is wrong. But that social knowledge nevertheless accounts for a large fraction of what we (think we) know, is clear.

Allowing for this social knowledge, and also for the default-like responses that seem to account for cognitive illusions like the chips puzzle, we can distinguish among three sources of intuitions a person feels confident about: direct experience, social experience (what "everyone knows"), and also—and most surprising, so that without the evidence that can be gleaned from cognitive illusions, we would never suppose it so—confidence that reflects merely well-entrenched defaults or otherwise logically weak but cognitively strong responses to unfamiliar contexts.[1]

The difficulty of challenging direct or social knowledge will naturally be greater when it happens to reinforce a default, and easier when it does not. But we can reasonably suppose a pecking order among these three classes: social knowledge should usually dominate a default or a distantly anchored anchor-and-adjust response; direct experience should dominate social knowledge. The kind of expert/lay conflicts that are our major concern then are most likely when direct experience is absent (so that both the danger and the costs of moderating danger in a situation are off-screen in terms of ordinary experience), but social knowledge is such that "everyone knows" that the risk is worrisome. But costs are too thinly spread or remote to envision an effect on "someone like me." So as just sketched (5.10) danger is on-screen and costs off-screen, which defines the risk matrix case where the "better safe than sorry" default is what makes sense to a person.

Particularly stubborn cases can be expected when "everyone knows" a substance is not only dangerous, but guilty (hence concern about fairness); or dangerous and inconsistent with the status quo (loss framing). But, as discussed in 4.12, even if neither adverse framing nor an adverse sense of fairness are clearly cued, both are likely to be seen anyway if fungibility is missing. A person is primed to pick up the fairness and framing sense that sits comfortably with the "better safe" intuitions that go with an awareness of danger and not of costs.

These responses will be very hard to counter by mere evidence and argument. Mere evidence or argument that conflicts with clear intuition, as could be seen in a situation as logically simple as the chips puzzle, is hard to believe, or hard to see as relevant. On the

---

1. I will refer mainly to defaults, but I mean also to include remote anchor-and-adjust responses, which sometimes yield intuitions that turn out to be brilliantly insightful and sometimes turn out to be illusory. For remotely anchored intuitions (remote, i.e., from the actual situation being faced) and for defaults, a person will ordinarily have no awareness of where the intuition comes from. And what comes to mind if a person is challenged to defend the intuition may have nothing to do with what turns out to prompt the intuition if a controlled experiment can be arranged (5.3).

other hand, if *direct experience* somehow pushes the costs of precautions on-screen, then (on the argument) we will now have a sense of fungibility instead of an unambiguous "better safe" intuition. A person would then be prompted to consider arguments that the risk may not be serious enough to warrant the costs of further precautions.

But if the case is one where in fact the arguments for attending to opportunities foregone are strong—which is to say, in just the cases marked by strong expert/lay controversy—an infusion of direct experience ought to be capable of producing striking shifts in intuition. For now (with the shift from cell 2 to cell 1 of the risk matrix) a person becomes open to evidence and argument, and especially so, of course, to strong evidence and argument. In terms of the earlier "twoness" argument, we ought to find cases where the strong lay sense of danger jumps to seeing that same risk as trivial.

And then, framing and fairness are likely to also prove susceptible to gestalt-like shifts, since if in a case where the latter mainly reflect network interactions (rather than being themselves well-anchored in experience) those interactions (once opportunity as well as danger, hence fungibility, is sensed) can more comfortably go the other way. These shifts would yield switches of fairness and framing that now make a bit more boldness seem also a bit more fair.

Finally, if something like the story just sketched is right, then we ought to be able to find clear examples by looking for what I will call *leverage cases*. Leverage cases occur when two risk situations are closely related. Ideally, the cases are exactly the same except for some particular aspect that injects an element of direct or social awareness of opportunity foregone where, in the alternative situation, that had been missing. And converse cases can be expected, where some novel aspect of direct or social experience prompts attention to danger where it was the danger side of the balance that had been missed.[2]

For such leverage cases, we could look for the effect of just

2. Our main concern has been with the polarized cases where lay intuition is in cell 2; only passing reference has been made to the converse cases. Nevertheless, as I have mentioned (4.2 and later), the argument, if correct, should be able to handle both, as well as the far more common cases where there is no polarized expert/lay conflict in either direction.

the feature that varies. In principle we might find leverage cases where it is framing or fairness intuitions that are changed by some infusion of direct experience. But on the "subtle risks" argument (4.12), such cases should be uncommon. Rather, we expect, fungibility (or its absence) will almost always govern the overall result. Given fungibility, people are ordinarily open to expert advice, which, for the kind of cases that mainly concern us here, would counsel against overreaction to risk. Then framing and fairness intuitions are likely to be pulled along. So if that argument here is right, we should expect to see striking reversals of intuition, not only about danger but also about framing and fairness. These shifts would follow some input from direct experience or perhaps some (ordinarily much slower) shift in social knowledge that pushes on-screen some opportunities foregone by extreme caution. And indeed it turns out to be easy to come up with examples of that.

6.2

In September 1993 the opening of the public schools in New York City was suspended in response to concern over asbestos. When the schools were closed, there was overwhelming popular sentiment for that. But the overwhelming view among people with experience to judge the seriousness of the threat was very different. The experts believed that the asbestos risk was not remotely serious enough to make it reasonable to think that the closing would benefit the children on whose behalf it was mandated. But there was not much trust in that expert assurance. Within three weeks, however, popular sentiment was overwhelmingly reversed, and it soon become difficult to find anyone who readily recalled being in favor of closing the schools in the first place.

So here the risk remained the same for the two risk situations (whatever the asbestos risk was when the schools were shut, it was essentially the same risk three weeks later).[3] But now the advantages of accepting the risk (costs avoided by not having to

---

3. Since some inspections were completed during the period, the risk was not *exactly* the same.

close the schools) became a matter of direct experience. Once the schools were closed, fungibility could no longer be escaped. Everyone with children now directly experienced costs to closing the schools, and the simple arguments that previously seemed ineffective turned out to be things ordinary people could understand after all: for example, that a child playing in the streets was not in a no-risk situation compared to a child attending a school that entailed some risk of exposure to asbestos.

Thus people no longer behaved as if they could not understand the possibility that a risk might be more than zero but still so small that a reasonable person would treat it as negligible: here a risk that (if in fact encountered by some child) could scarcely amount to much more than a 1 in a 1,000,000 chance of an eventual cancer. That would be in addition to the 1 in 5 risk that everyone runs of eventually getting cancer, no matter how cautious they or their government might be, and of course the 5 out of 5 chance that, if a person does not die of cancer, she will die of something else.

This New York case provides a striking illustration of how quickly direct experience can turn a situation in which "everyone knows" that "better safe than sorry" is the right response into a situation in which "everyone knows" that we better think about the trade-offs, listen to what experts can tell us, and, in general, respond to the fact of fungibility.

But several other instructive features are also apparent. Asbestos itself had become firmly perceived as a substance "everyone knows" is very worrisome by high rates of lung cancer among World War II shipyard workers that appeared several decades later. That has proved sufficient to make perceptions of exposure to asbestos firmly anchored in danger; it has also made it easy to color those perceptions with guilt, however remote the current context from the context that anchors those perceptions.

In New York, the adverse fairness perception was reinforced from another angle. The crisis mood developed from scandal about corruption and incompetence in carrying out Congressionally mandated inspections, where the Congressional mandate itself reflected the same distantly anchored perceptions that governed the New York public response. As very often happens, strong public concern prompts strong Congressional or agency mandates, which are then read to justify public concern that the situation really is serious (if the problem is not serious, why would Congress act, why would the agency commission expensive studies and monitoring, etc.). For asbestos there was both a past history and current behavior that together provided a striking case of a substance stigmatized by both danger and guilt.

Nevertheless, with direct experience prompting fungibility, that concern melted quickly. Questions of fairness did not disappear. But what was salient for that concern shifted radically. People saw children as victims of panicky officials who imprudently closed the schools, not as victims of corrupt and careless contractors. Asbestos in the schools was no longer seen as a newly discovered risk (imposing losses from what had been taken to be the status quo) but a risk that had always been there as part of the status quo.[4]

The Congressional concern which mandated school inspections itself reflected the substance/dose effect (4.6). "Everyone knows" asbestos is bad, and asbestos in schools (so any problem peculiarly affects children) easily prompts fairness concerns with children as victims. There was much talk of absolute removal. On the other hand, trying to remove all asbestos everywhere would force fungibility: the costs would be sufficiently enormous that it would quickly become apparent that attention had to be paid to where the money would come from. So politically alert advocates of asbestos removal pushed for the strongest position they could realistically aim for, which was extremely cautious standards, not absolute removal.

Twoness then (and characteristically) entered again. Once

---

4. The stigma rooted in the shipyard exposures was not only because of sympathy for unprotected workers, but because of extensive press attention to the failure of management to warn workers even after evidence of health problems began to come in. There are things to be said for mitigating the situation. The exposures were mainly during wartime, when there were severe fire risks on ships, so that it is likely that many more sailors' lives (far more at risk, anyway, than workers on the home front) were saved than workers' lives lost. Knowledge of the risk that is now well established was not so then. Safety practices that are now routine (masks) were not so then. But nothing in this, on the record, significantly offsets the cognitive salience of the adverse elements.

chosen, the standard became a bright line between safety and reck-lessness. Below the standard the risk was negligible; above the standard—even barely above the standard for a short time—was a risk that might kill a child, or at least might kill a former child if a cancer developed decades later. All this yielded serious concern about even a possibility of asbestos exposure in any degree above the standard. Not only was the calculated risk exceedingly small (on the order of one in a million), but calculated, as is customary in these matters, in an extremely cautious way. And asbestos in a school, after all, has some chance of saving a child's life in the event of fire. Nevertheless, until fungibility entered, any failure to comply seemed absolutely dangerous, compounded in New York by corruption and neglect in handling the mandated inspec-tions. All this prompted a strong sense of an emergency that demanded a strong response. But that lasted only until direct expe-rience prompted awareness of costs sharp enough not to be missed.

Reviewing: we start from a particularly strong "better safe than sorry" default from the mere mention of asbestos (which "everyone knows" is dangerous and everyone feels is tainted by guilt), which can be seen as a new risk (loss framing) since al-though the asbestos was there a long time and was intended to be and, on net in fact probably was, protective (as a fire retardant), the lay sense is of a loss framing focused on the newly attended risk. The perceived status quo (what felt like the status quo) was a situation with no worry about asbestos, which was not a situa-tion with no asbestos. Or the cognitively effective status quo was what the situation should have been if the inspections had been carried out correctly. Framing can go either way (the risk has been around for decades, but should have been attended to). So if dan-ger is on-screen, but costs not so, framing easily falls in line with a "better safe than sorry" sense of the situation. But direct experience with the consequences of closing the schools put costs as well as danger on-screen. And that (introducing fungibility) made it seem obvious to seek expert advice. Very soon the sense of where danger might lie reversed, and framing and fairness intu-itions reversed as well.

The completeness of the reversal of intuition that followed awareness of fungibility needs a little emphasis. For most people, the risk from asbestos soon no longer seemed important at all. It was not that people now felt that, as bad as the asbestos risk was, it had to be accepted given the even greater costs and risks of keeping the schools closed. Rather, over a very few weeks the risk of asbestos simply disappeared. Expert arguments about the minimal character of the risk took hold as what "everyone knows" about this risk replacing what everybody had known only a short time earlier. Once the key element of fungibility flips other elements of the network also easily flip: things remain as black and white as before, but what was black becomes white and what was white becomes black.

6.3

Comparable effects show up on many other matters. Reports of contaminants like lead or arsenic readily prompt strong "better safe" responses. If all we see in a context is that it involves exposure to something "everyone knows" is a poison (lead, arsenic), then of course the default intuition prompted can only be "better safe." But in some communities, like Aspen, Colorado, and Triumph, Idaho (not far from Sun Valley), which are built on mine tailings contaminated by heavy metal far exceeding the EPA's action levels, strong local resistance developed toward EPA clean-ups. Responding to this, state governors, who more commonly are demanding that the EPA act to protect their citizens, were in these cases demanding that the EPA stop persecuting their citizens.

The difference in response of similar populations to similar risks is very directly a matter of fungibility. If the community faces severe disruption, people soon shift from a polarized (cell 2) "better safe" first response, and prove willing to listen to what experts can tell them about the risk they face. But if an outside agent is to bear the costs, and no gross disruption of the community is involved (the usual case, e.g., for concern over possible seepage from a dump), then costs are off-screen, and "better safe than sorry" overwhelms argument and evidence. Local people are outraged. They do not trust anyone who tells them that the risk is trivial by their own normal standards of what is worth worrying about. They gather cases of cancer to show the terrible effects, which is never very hard to do, since deaths from cancer are common and, barring immortality, probably must remain so.[5]

Framing and fairness shifts again can be seen, as in New York. But here there is more opportunity to distinguish between local and more remote views. From the local view, fungibility enters through disruption of the community (for a very large community, that is what happened in New York City after the schools had been closed a while). For the wider society, fungibility is logically always present: citizens at large—in particular, citizens as taxpayers—will have to pay. But cognitively such costs are diffused thinly across citizens. Unless the costs come to take a noticeable fraction of the total budget, they only obscurely relate to what happens when taxes have to be paid by an individual. Empathy makes it easy for us to envision possible costs to some individual victim (e.g., dying of cancer). Nothing makes it comparably easy for a person to envision the costs to a person (as taxpayer) of using resources for precautions, or the costs to a person (as resident of a community) of forgoing opportunities, until the costs of forgone opportunities grow so large that it becomes obvious that projects that interest a person will be lost or that taxes will have to go up to accommodate the precautions.

The main counterexamples here are instructive: a hammer for which the U.S. Air Force was billed $500 has becomes part of American folklore, as has a scandal focused on banking arrangements at the House of Representatives. Both readily related to the direct experience of ordinary people, though the first was a quirk of defense contracting procedures and the second involved no loss to taxpayers. But an ordinary person knows he is being robbed if he is billed $500 for a hammer, and he knows that he would be in trouble if he repeatedly wrote checks that exceeded what he

5. What about cases where there clearly is an excess incidence of some kind? In both Woburn, Mass., e.g., or near the nuclear installations at Sellafield, England, there was a clearly excessive number of childhood leukemias. In Woburn, as I write, industrial pollution indeed proved to be the cause, though not from the source originally suspected. At Sellafield, the nuclear processing plant proved to be a plausible suspect that turned out to be innocent. See Neel (1994); see also Hattchouel et al. (1995) and Kinlen et al. (1995).

had in the bank. But a billion dollars misused in a way that does not relate to what a person can readily grasp in terms of his own experience usually causes no great fuss. So the counterexamples (where the public is sensitive to socially small costs—in these examples, even socially microscopic costs) fit the general cognitive story here very well.

In New York, the belief seems to have been that the schools would be closed for a short period, properly inspected, and then they would be reopened with the asbestos risk gone. This scenario quickly proved to be unrealistic. Indeed it could not be realistic, since if the inspections were done quickly that essentially guaranteed that news reports would appear complaining that they were being done sloppily. But at the time of the crisis, there was not much willingness to listen to an argument that conflicted with what "everyone knows" was the right thing to do.

In Aspen and Triumph, however, the EPA's plans for removing the mine tailings plainly would require disrupting homes built upon those tailings. Even if residents were fully compensated at market prices it was unlikely that the modest income housing involved would ever return to these sites, or to anyplace nearby. For land anywhere in the area had become worth much more as sites for vacation homes for people far more rich and celebrated. So residents could not fail to see that there would be unmistakable pain well beyond any financial compensation to those the cleanup was supposed to be helping. Since the prospective pain was severe, residents in both communities were very willing to listen to expert doubts that any substantial risk was actually involved (e.g., listen to evidence from blood testing that their children did not show the high lead burdens that EPA's ultracautious models predicted) or to arguments that the heavy trucking required for the EPA plan was likely to have more tragic consequences than the cleanup could plausibly prevent.

Consequently, what at first was taken as a fearsome risk, with residents as victims of irresponsible early mining practices, now was seen as trivial. The EPA officials, initially perceived as defenders of victims, were soon seen as the persecutors of victims. The lack of salience of costs to be borne by outsiders for the cleanup was replaced by the salience of the loss of their community to the residents. Even more clearly than in New York, once fungibility entered, framing reversed. The contamination was now seen as old, not the discovery of contamination far above EPA action levels as new. The risk itself came to be seen as obviously trivial instead of obviously frightening. And fairness intuitions also reversed. Residents saw themselves as the victims of bureaucratic and environmental zealots rather than the victims of careless mine operators. As with the New York case, we see both the strong effect of fungibility and the accompanying shifts of gestalt that bring perceptions on fairness and framing in line with a view that now sees the risk as not worth worrying about.

6.4

A variant of the Aspen lead case is the contamination of the area around the Hanford, Washington, nuclear site with arsenic. Soil levels at some locations are a thousand times the EPA's action levels. Other than a handful of small, high-security parcels within the Hanford exclusion zone, by the very standards that rationalize the spending of $2 billion a year on the Hanford cleanup, these arsenic levels may be (by EPA standards) the most dangerous contamination in the area. Yet the government has attracted thousands of families, many with small children, into the area. So although the hazard is natural, the exposure of people to the hazard often comes from a government action.

But few people seem to care. And in terms of the risk matrix the reason is straightforward. Arsenic is element 33 on the periodic table. Like any other element it can be found everywhere, but like gold and other uncommon elements only a relatively few places have concentrations high enough to stand out. When a person first hears of the high arsenic concentrations near Hanford, it comes very naturally to suppose that somehow Hanford produced the contamination. But in fact high arsenic levels are common across a large area of southeastern Washington. Some of this may be attributed to mining and smelting operations. Mostly, though, it was there not only long before nuclear energy, and before mining began, but before the existence of human beings.

But once this is known, "better safe than sorry" intuitions

dissipate. If many tracts of southeastern Washington State and neighboring Oregon have this contamination, for which no one but nature can be blamed and which would obviously involve bizarre costs and difficulties to remove, then things are stark enough for fungibility to enter. As at Aspen and Triumph, by and large people are then quite willing to listen to what experts have to say about whether there is any evidence of arsenic poisoning in the area, despite the demonstrable excesses in terms of EPA standards.

But conditions that prompt awareness of fungibility for one person need not do that for another, since what is direct experience for one person may be irrelevant or opaque to another. People who live in a community may feel satisfied that they are not exposed to some unreasonable risk, but outsiders may be sure they are mistaken. So there has been general support for a nuclear waste site (i.e., the WIPP—the "waste isolation pilot plant") in Carlsbad, New Mexico. Ongoing activity there provides many economic benefits (hence fungibility). And fungibility would also be favored by a wholly different kind of awareness of costs. In Carlsbad, friends and neighbors describe elaborate precautions already in place and complain about the unfairness of outsiders who exaggerate the risk or invent risks that are not there at all.

Coming to believe you cannot trust your friends and neighbors is painful. So a person in Carlsbad is more willing to listen to what people who work on a project in Carlsbad have to say than would a similar person in Santa Fe, several hundred miles away. There are no benefits in Santa Fe from economic activity in Carlsbad, and no pain in distrusting people you have never met. Opposition to burying nuclear waste near Carlsbad has been intense in Santa Fe, where nothing from direct experience challenges "better safe than sorry." Even if the case for WIPP were seen as apparently reasonable, it is always possible that if it is studied some more—rather than allowing it to go forward— something bad will turn up.[6]

6. Opposition in Santa Fe was made more intense by the fact that shipments of waste from Los Alamos to Carlsbad would pass through or at least skirt Santa Fe. But those shipments would not begin until after years of ship-

6.5

In general, even a remote chance of exposure to radiation from nuclear waste causes concern vastly larger than a certainty of comparable exposure to radiation from flying in an airplane (which necessarily means exposure to more cosmic rays) or vacationing in the Rocky Mountains (which feature high altitude and also uranium deposits), or walking past a display of smoke detectors while shopping in a hardware store.

Many people are extremely concerned about radiation waste being stored anywhere in their state, but few people need anything more than a quick reassurance from their doctor to agree to have that same material injected into their body as part of a medical procedure. We do not hesitate to take that expert advice, though if the same doctor told us that a low-level waste dump somewhere in their state created no nontrivial risk, we might find that hard to believe. But fungibility is immediate in the first case: a person wants to be cured. Fungibility is hard to sense in the second, since the costs of precautions are hard to see in terms of what a person has experience with as an individual. A major source of low level waste is medicine. But it is not easy to see that if dealing with waste is made difficult, some people (perhaps even the very person now quite willing to undergo a test using radioactive isotopes) might not get a potentially life-saving test.

Compare the situation to a proposal to ban individuals infected with HIV from airplanes on the grounds that shared lavatories—even shared armrests—expose uninfected passengers to a risk of catching AIDS. This risk is trivial, but it is not zero, and in fact could hardly fail to be larger than some risks from radioactive

---

ments from other points, so residents would have an enormous backlog of data to either be reassured by or to demand severe precautions when that time came. The dominant factor seems to me more plausibly a sense of concern and responsibility for what happens in a person's state, where a concerned citizen is likely to more easily feel some responsibility and also can make more of a difference than she can on a project in another state. A similar pattern of especially severe opposition from a distant city but within the state appears in the particularly severe opposition to the Yucca Mountain nuclear waste site from Reno, several hundred miles to the north.

waste or asbestos or dioxin that have prompted intense concern. But here it is easy to envision the costs that such a precaution would impose on people who are already tragically burdened. So almost everyone would respond to such a proposal with distaste for the people who propose it, and a feeling would be expressed that it is stupid or neurotic to worry about such a trivial risk.

6.6

Are there cases where fungibility *follows*, rather than leads, framing and fairness? I have suggested that clear cases of that sort would be hard to find. But treating two of the three Fs as merely contingent on fungibility pushes things further than what we can see warrants, and also further than the argument I am making requires. As this is written, there is lively controversy over government-funded medical experiments involving the deliberate exposure of patients to radiation. And it is apparent that experiments done under military auspices are regarded with much more suspicion than experiments that, on the face of things, are very similar but were sponsored by the National Institutes of Health (NIH). Fairness considerations here surely may be inhibiting fungibility at least as much as the absence of fungibility is promoting intuitions of unfairness. If it is easier to suspect a military than a civilian medical experiment of unfairly harming people, then that propensity presumably would make it also harder to see anything worth taking seriously in justifications of a military, as opposed to a civilian, experiment.

A person can more readily—in time of peace, at any rate—see participants in Department of Defense experiments as victims because it is easier to attribute guilt to the military than to civilian medical research. And I have already mentioned substances (dioxin and asbestos being salient examples) that come into play already severely tinged with guilt before any further context is supplied. Radioactive waste, even when it produced by civilian medicine, seems to be contaminated by association with military waste. On the other hand, waste, by definition, is prime material for strong "better safe than sorry" intuitions. What is it good for? So there would be a direct influence discouraging fungibility from the mere fact that what is at issue is waste, but also some influence discouraging a sense of fungibility from the propensity to see this *particular* waste—because of its association in people's minds with military activity—as intrinsically guilty.

Overall, though, so far as I have seen, even hard cases turn, at bottom, on fungibility. As illustrated by the New York asbestos case, direct experience can prompt fungibility even in relation to a conspicuously guilty substance, and once in place fungibility then can reverse fairness and framing.

6.7

A curious extension of danger (and sometimes guilt) by psychic contamination comes from controversy over *irradiation* of foods. This is a technology for destroying microbes contaminating foods. There is no reasonable doubt it would save lives. And although radioactive isotopes are used in the process, no residual induced radioactivity is involved. So here the product is not radioactive at all: it has merely been exposed to radioactivity, like the folktale of the chicken broth so thin it was made by exposure to the shadow of a chicken.

But fungibility comes hard here. In ordinary experience the food supply is safe (sickness occurs, but too rarely for an individual to come to see the risk as personally significant). And it happens that irradiation is more expensive than older techniques, so that foods processed in this way would be safer, but by a very small margin on the scale of all risks a person encounters. Since irradiation is more expensive than chemical alternatives, food processors are not motivated to favor it, while consumers (from the link to radiation) are prompted to "better safe than sorry" worries. The network effects on framing then favor seeing the novel technology as a risky departure from the status quo, although in other contexts, where fungibility could not be missed, the same situation could be seen as a better way to carry out what is already part of the status quo: procedures necessary to destroy microbes that might contaminate the food supply. In one case, consumers of irradiated food are seen as victims of an experimental technology. But given fungibility, the situation would almost certainly be seen as one where victims of food poisoning are being created by a failure to make use of the best available technology.

6.8

Consider another case where, on the face of things, guilt seems to be an important element. Aflatoxin and dioxin are alike in that both are unintended contaminants that no one deliberately makes. Between the two, aflatoxin is on balance the more important risk, but dioxin is by far what we worry about more. This is very marked, even though children are a particular target for aflatoxin, since a main source is peanut butter.[7]

But aflatoxin is guilt-free, and dioxin is peculiarly encumbered with guilt. Concern about dioxin originally arose in connection with the heavy use of defoliants in Vietnam. American and Australian veterans remain convinced that the presence of dioxin as a trace contaminant in the defoliants created serious health effects, though elaborate studies in both countries have failed to find much evidence for such effects.[8] Aflatoxin comes from a mold that contaminates peanuts and grain and reaches people when they consume these foods. Since it is natural and no one knows how it can be entirely avoided, it is less easily tinged with guilt and is easier to see as a risk that has always been around.

Current exposures to dioxin come from trace amounts released into the atmosphere, mainly after being created by combustion of industrial and medical wastes; apparently some dioxin also

---

7. Aflatoxin is far more plausibly the more serious cause of cancer to humans, essentially because at any given (but ordinarily very small for both) fraction of the dose at which it is a demonstrated carcinogen, aflatoxin is more commonly encountered in the environment than dioxin. But also there is evidence (in particular from follow-up studies of cancer incidence following a major industrial accident at Seveso, Italy, in 1978) that low concentrations of dioxin somehow inhibit certain types of cancer, as well as encouraging others. On net there was an overall lowering of cancer incidence. For a broad survey of the dioxin case, see Gough (1993).

8. As usual in historical or scientific disputes (not just in environmental disputes), the situation is not that there are no studies that support the contrary view, but that such studies as do seem to support claims of harm are early or crude or (mostly) both. The more careful and sophisticated, the less likely it has been to find effects.

is created by forest fires. Logically, current exposure has nothing to do with feared harm to Vietnam veterans, but, as with asbestos, guilt by association sticks. One might suppose that since children are the most enthusiastic consumers of peanut butter, there would be a strong fairness cue to worry about aflatoxin. After all, even if occasional mold in stored peanuts is natural and inevitable, making peanuts into peanut butter and feeding it to children is not. Noting how dioxin has come to be stigmatized by guilt seems essential to understanding why it has gotten special attention. But it is not so clear why trace dioxin unconnected with what anyone claims to be inappropriate activities prompts much sharper fairness concerns than aflatoxin.

But if we look to fungibility, the reason for the contrast is not hard to see. Children like peanut butter a lot, and adults recall liking peanut butter a lot and sometimes still do. So giving up peanut butter is something that readily prompts fungibility. A person wants to know whether the risk is serious enough to warrant giving up something as nice as peanut butter, and so she is quite willing to listen to arguments about how significant the risk might be weighed against the health merits of peanuts. But there is no such easy route to fungibility for dioxin. It is just an unwanted by-product. The costs of extreme precautions run to some billions of dollars, but spread sufficiently thinly across the economy that few people can see any effect due to the precautions proposed. An instructive contrast here is to the AIDS risk mentioned at the end of (6.5).

6.9

Another instructively contrasting pairing is genetic engineering and nuclear waste. Each concerns danger linked to exotic technology, and both were matters of conspicuous concern in the 1970s. But despite periodic worrying news stories and the energetic efforts of activists to promote concern, for genetic engineering reassuring the public has turned out to be fairly easy. In contrast, concern about nuclear waste has not faded at all. So we want to look for an explanation of this contrast.

For both, there is a local and a wider context. Some local

community has to be the host for a site or firm or field experiment (for genetic engineering) or as a storage site (for nuclear waste). But even if that is available, opposition can arise from surrounding communities or from state or national constraints. I will start by sketching the situation with respect to a nuclear waste site, and then contrast that with the happier (for industry, and perhaps also for common sense) outcomes that prevail for genetic engineering.

If you point to some community at random, it will almost surely be opposed to becoming the home of a nuclear waste site. So the situation will easily be very difficult if a site has been designated in advance, and the effort has to be to prompt fungibility in that designated community. But if there is an opportunity for informal discussion leaders from many diverse communities, each aware that similar informal discussions are being held with people in other communities, the dynamics of bargaining and negotiation and eventually public discussion can be very different. There are many individual communities, only one of which needs to be willing to accept the site. And the economics of the situation will provide bargaining room for a generous set of incentives to encourage any community that can be interested.

But if the expert consensus on risks here is right, once fungibility is sensed in some community (once there is a sense that there may be a trade-off worth considering) a strong case can then be made that in fact risks are minimal and benefits substantial. So we can expect (on the risk matrix argument) that if only local opposition needs to be considered, the problem of locating a site should not be overwhelming. Indeed, possible advantages to a community may be apparent enough that the community itself takes the initiative in inviting attention as a possible site, as indeed occurred at Carlsbad. Then fungibility is at hand even before any discussion of a possible arrangement starts.

But suppose that a potentially willing community indeed has been found. A further critical matter will be whether that community would be allowed to move ahead. There might be opposition from surrounding communities (which cannot easily see any benefits) or from still wider opposition at state or national levels. The local/wider distinction that has already arisen in the comments on WIPP (6.4) then becomes very important and, in fact, has repeatedly proved crucial.

In general the social benefits of finding a site grow increasingly thinly spread and otherwise hard to perceive as the horizon is widened. And compensating benefits (employment opportunities, civic amenities from a larger tax base) dissipate quickly with distance. Opposition, consequently, typically reveals a volcano-shaped pattern. For a case where the local community is on balance willing to go ahead, there will still remain some local opposition. But opposition will grow as you move *away* from the local area, typically with a drop in concern only beyond some salient boundary, such as a county line or perhaps only at state boundaries. For nuclear waste and genetic engineering, opposition remains important even at the national level.

Now it makes sense in terms of the fungibility argument that for subtle risks—in particular for risks so subtle that predominant expert judgment is that they are scarcely risks at all—local opposition can sometimes be overcome more easily than more remote opposition. At the same time, and on the same evidence, a person somewhat removed from the site might see nothing to gain from whatever advantages would come with accepting the site. So with fungibility absent, it would be hard to satisfy that person that there is nothing to worry about. And if there might be something to worry about, then whatever it is might leak over to where I am, or be spread by risks of transportation, as in Santa Fe, several hundred miles from Carlsbad.

Further, a concerned citizen may feel some responsibility for what is going on in his country or her state. And without the spur of fungibility, a person will not easily discriminate between different levels of risk. Even empathy has inverted effects here. I think I know how I would feel about the danger, were I in that town. But I do not so easily see how I would feel about compensating advantages if I lived in that town. So in the end, I also do not really know how I would feel about the danger. A person who is insensitive to possible advantages, hence never feels a sense of fungibility, has no way to see that the danger itself might come to seem trivial, even if in the case at hand that would be a wholly reasonable assessment—even if, in fact, it is very reasonable to see the risk as zero. Such a person (lacking any sense of fungibility) readily comes to see the community willing to accept the site as in need of protection from their own naivety, or economic vulnerability; or that their children or future generations or some other victim needs that protection. Overall, then, the "volcano" effect, where concern is highest in peripheral locations, not in the immediate vicinity of a risk, is not hard to explain.

6.10

So far, however, almost everything should go the same way for genetic engineering. Of course, nuclear waste has the special burden of guilt by association with radioactive fallout. But genetic engineering has the special difficulty of association with a very long tradition of Frankenstein monsters, pacts with the devil to gain terrible and ultimately disastrous knowledge, meddling with what is natural. And a genetically engineered microbe leaking out of a laboratory surely seems at least as worrisome as a leak of radioactive material. The microbe might very soon multiply by billions, which even the most naive worrier about radioactive waste knows cannot happen to a bit of waste. The microbe can invade your body and multiply by billions right there.

But the genetic engineering facility has offsetting advantages, and large ones, though I want to argue that even they are not crucial. A dump is not a dump, in the sense that not all dumps look (or smell) alike. A nuclear waste site is a high-tech facility that does not look especially bad, does not smell, and indeed can be designed and landscaped to look quite attractive. Nevertheless, a genetic engineering facility does not have to contend with the label "dump," and hence it does not entail the stigma that goes with a community that is home to a dump, however nicely designed.

Nor will even a well-designed dump look like the elegantly designed and landscaped facilities that will be built by genetic engineering firms. Nor will a dump attract highly paid professionals who will live in the closely surrounding area, patronize upscale shops and restaurants, raise property values, provide welcome sorts of newcomers for the low-crime residential communities such people can afford to live in.[9] And the prosperous, well-educated communities likely to be of interest to the genetic engineering firm have a good deal more access to levers of power in the society than do the communities likely to be of interest as waste sites. They are also less vulnerable to claims that the residents are too naive, poor, and so on, to exercise good judgment about whether a risk would be reasonable. So, other things equal, the former, more easily than the latter, can answer concerns from the surrounding area.

Nevertheless, the most critical contrast between the nuclear waste and genetic engineering cases appears at the national level. Choices at the national level about regulations and related matters may set rules that make it easier to build opposition to a site; but sometimes the opposite is true when federal mandates preempt local areas from imposing their own standards. We want to look for something that would make the costs of inhibiting genetic engineering readily perceptible at the national level but leave the costs of inhibiting nuclear activities (by making it difficult or impossible to deal with the waste from such activities) almost invisible.

But that is not hard to do. On the nuclear side, difficulties in finding waste sites add to the costs of activities like nuclear medicine or nuclear power. They effectively stop some such activities by increasing the costs: activities at the margin are necessarily priced out. But so long as supplies of fossil fuel remain ample, power is available, and nuclear medicine is not enough more expensive or inhibited enough to make any difference noticeable to anyone but medical specialists. Someday that will no longer be true, at least with respect to power. For no one supposes that supplies of petroleum at something like current (real) prices are likely to last out the coming century, and few doubt that a massive shift to coal would involve far more environmental and health

9. On the other hand, that may only explain why communities that are willing to consider a nuclear waste site are not likely to be well-to-do residential communities, even if other considerations did not make a metropolitan area a poor prospect for a nuclear waste site anyway.

risks than moving towards nuclear power. So eventually fungibility will be forced on-screen at the national level, and at that point (if other circumstances have not already resolved the matter) there will be enough concern among enough of "the establishment" that national regulations and attitudes that currently make it hard for a willing community to become a waste site will move the other way. But that has not happened yet, since today the costs of crippling further expansion of nuclear power are not huge and challenging what "everyone knows" about the dangers of nuclear waste is not a high priority for enough people to push such policies through.

But for genetic engineering, the situation was and remains quite different. By the end of the 1970s it was quite obvious to almost any reasonably knowledgeable person that genetic engineering was going to be of vast economic and scientific importance in the twenty-first century, and that if the country chose to drive such activity abroad, the costs would eventually be very large. In contrast to nuclear power, which was not yet economically highly advantageous compared to fossil fuel plants (so there was no conspicuous loss that could be pointed to) genetic engineering could point to new medical achievements and to people whose lives were enormously helped by the products of this technology.

There is, of course, more to this issue. In Germany, restrictions on genetic engineering indeed have been crippling, at least into the mid-1990s,[10] which is tied on one side to the stigma of the Nazi experiments on human beings and on the other to the fact that Germany (a country with no petroleum resources) has strong motivation to move ahead on nuclear power, so that genetic engineering was an easier "green" target there. And German genetic engineering investment could be moved to other countries (no such inhibitions prevail in France or Switzerland) without moving very far.

But none of this is inconsistent with the main argument here, which leads us to expect that the critical differences, as we look across cases (of different risks, in different countries, etc.) will

10. See *Nature* (18 May 1995, 175) for a report on changing German attitudes, as would be expected sooner or later on the risk matrix argument.

above all be differences in how readily fungibility can be prompted (now at the level of national consequences) in one situation as against another.

6.11

Does the sample of cases in this chapter prove that the risk matrix story I have been urging is clearly right? That would be expecting too much. There is always room for dispute over the interpretation of any case on the table. And even if you are satisfied that indeed the cases support the reading I have given them, they still are only a small sample of all such cases. But collectively (it seems to me) the cases provide very good support for the plausibility of the risk matrix story. I hope they are sufficient to challenge a reader who believes there must be plenty of counterexamples to see if he can think of one. In any case, since the analysis here is different from the usual story, we can expect it to lead to different prescriptions of what might help resolve expert/lay conflicts. In the remaining chapters, I will try to show how that might work.

Two Modest Proposals:
Some Background

7

Suppose the analysis to this point is at least roughly on target. What does it suggest? The way to start is by considering various familiar efforts and see why, in terms of the risk matrix, they are difficult to make effective, and what the risk matrix can suggest about what might be done. The familiar proposals I will review deal in part with the *efficiency* of social choices about risk, and also (but other proposals) with *equity* in the process of making those choices. The two modest but in some way novel proposals I will comment on (in chaps. 8 and 9, respectively) draw on aspects of both efficiency and equity. Both are tied to the problem of public trust.

Since the public manifestly does not believe government assurances about certain risks, on what I have been calling the "usual story" an essential component of resolving the difficulties must be building trust.[1] Successive EPA administrators and, even more conspicuously, successive secretaries of energy, along with many less beleaguered agency heads, have mounted ever more elaborate efforts to build trust by improved public access to information and by greater public participation in environmental choices. And these efforts have been substantial, not mere talk. But a gain in public trust is not easy to see, which gives a certain face validity

1. See, e.g., the analysis of the nuclear waste problem in Slovic, Flynn, and Layman (1991).

to the possibility that building trust—through greater efforts devoted to openness and to listening to what the public has to say—is not the remedy that it has been so often claimed to be.

But what goes wrong is not a deep mystery. Building trust is a good thing. Yet it takes a measure of naivete to suppose that absolute candor by a public official is a formula that will reliably build trust. Effective leadership indeed requires trust. But it is naive to imagine that successful political leadership has ever been marked by absolute candor. Consequently it is also naive to think that lack of trust is really explained by supposing that any departure—indeed, as the argument is commonly made, even merely suspected departure—from total candor is what accounts for a public lack of trust. The public itself is not that naive.

One merit of the risk matrix view is that it accounts for the manifest lack of trust in usually accepted voices of authority (where expert/lay conflicts are in play) without embracing a view that violates an obvious fact about politics in every time and place we know anything about. The point is not that a reputation for candor is unimportant in politics, or that a reputation for candor can be sustained in a democratic society without a measure of actual candor. But in politics, unqualified candor is an impossibility (or a disaster, if actually attempted). If adequate trust for authorities to carry out their responsibilities requires meeting that impossible standard, then things can only go badly. On the view here, sufficient trust in fact would return, given attainable (not imaginary) levels of official competence and candor once costs of stalemate (or of overreaction to minimal risks) become sufficiently clear. This is just the fungibility story I have been stressing for many pages now. On that view, what will build trust is not so much explicit trust building, but changes in *process* that make the costs of single-minded "better safe than sorry" responses more visible.

But given wide belief in the usual story, trust-building activity (whether or not it actually results in building trust) is attractive and indeed indispensable to harried officials. In the short run, it reaps praise and avoids criticism for failing to try harder to establish trust. Further, it excuses lack of action on actual problems. So it is not surprising that trust building remains something enthusiastically spoken of, even in the face of the manifest failure of much talk and effort in that direction to actually produce more trust. If the risk matrix argument is close to right, it would not be realistic to expect that a person firmly seeing "better safe than sorry" (firmly polarized in cell 2 of the risk matrix) will somehow come to trust someone who is telling him something that he finds he does not believe (2.6).

7.2

Trust building has not built much trust. But neither have efforts to push efficiency gotten much more efficiency. I will (in chap. 9) be commenting on how more concern for efficiency—which on the argument here depends heavily on making fungibility visible—might be built into public participation efforts. But that requires efforts to promote efficiency that do not somehow seem to seek efficiency by sacrificing equity. Before going further, though, it will be helpful to have a quick review of the most common proposals for improving efficiency. One familiar idea is to find a way to define *de minimis* standards for risk. This idea appeals to the very old common law principal that the law need not concern itself with trivial harms. In terms of economic rationality, this translates into an injunction to avoid adjudication of matters so inconsequential that the costs of adjudication (and, for regulation, necessarily including also the costs that would be involved in developing and legislating and litigating and enforcing the regulations) would outrun the damage that might be prevented.

A second efficiency theme, and a more fundamental one, concerns the importance of attention to diminishing marginal returns, or what Breyer calls the "last 10%" problem, or which also might be called the "how much is enough" problem. Other things equal, as stricter and stricter controls are considered, either the cost of a further increment of control escalates, or what might be gained decreases, and commonly both. So we are paying more and more to get less and less. But fundamental though such concerns are for effective allocation of resources, it has proved very difficult to reliably attract attention to diminishing marginal returns. Several of the most important pieces of environmental legislation on their face forbid the explicit attention to marginal costs and benefits that is required to get at questions of efficiency.

A third persistent effort is to present various risks in some side-by-side way that might put intuitions about one risk in perspective with other intuitions about risk. The presumption is that, if we are treating some risks as far more serious than might be supposed from their actuarial significance, and others far less seriously, then we at least ought to notice the difference. Perennial attempts are made to build such comparisons into regulatory management.

In chapter 8, I will take up a "do no harm" proposal that aims to complement these familiar proposals for promoting efficiency. But the proposal, though directly concerned with promoting fungibility, hence efficiency, seeks to do that primarily by appealing to intuitions about fairness. Parallel to that, a proposal about fairness and public participation, which will be the main focus in chapter 9, is mainly concerned with efficiency (again tied to awareness of fungibility). So both proposals straddle the efficiency/ equity division, as the risk matrix analysis suggests should be done. For the main result of that analysis is that fungibility will often be hard to prompt for subtle risks, and that will be true in particular so long as the fairness gestalt is such that efforts to push efficiency are perceived as just devices to paper over the exploitation of victims.

What I have called network interactions then powerfully lock in a "better safe than sorry" default (cell 2 of the risk matrix) that makes analysis seem irrelevant and makes *de minimis* claims and arguments about diminishing marginal returns seem merely about sacrificing a few victims to save the rest of us some money we will never really see anyway. Why would anyone be interested in that? And risk ladder comparisons take on that aspect as soon as they are called upon to rationalize leaving some risk uncorrected.

Hence the usual reforms prove hard to sell, since they are seen as protecting inequity. But without efficiency reforms, efforts to show that the process is fair (by enhanced public participation and open information) are likely to be eventually denounced as just a trick, or to lead only to frustration. For eventually it will become apparent that resources are not unlimited, so that funds to provide as much response to "better safe" concerns as participants think that they are being promised are not going to be found.

7.3

Suppose you are a skier (or if not, think of some suitable example of something you do know from firsthand experience), and I ask whether skiing is safe. You are not at all likely to have an intuitive response of simply no (or presumably you would not be a skier). Nor are you likely to have a response that is simply yes, since you are no doubt familiar with situations in which a person could easily be injured, or you have seen someone injured, or you have even been injured yourself. So your response is probably going to be something nuanced, such as, "Yes, if you exercise due care." Even if your initial intuition was just yes, it will be easy to nudge you toward the nuanced answer. But if I ask you about a hazard outside direct experience, and if the question is about some substance that "everyone knows" is dangerous, then a nuanced intuition will not come easily. That is likely to prove so even if you are inclined to try, as (on a far simpler matter) most of us find it hard to feel comfortable with the 2/3 answer to the chips puzzle of chapter 3.

But as the cases examined in chapter 6 illustrate, such responses, though not easily overcome, can in fact be overcome if an effective rival intuition is prompted, so that a person comes to feel the possibility that choices could be too cautious as well as the possibility that they could be too bold. In the language I have mainly used, circumstances (but rarely mere logic) can sometimes prompt a sense of fungibility, at which point logic does become relevant.

Then it turns out, as illustrated in the cases examined in chapter 6, that it is not really a matter of people being immune in some context-free sense to the sorts of arguments urged by the reformers, but of people being immune if their sense of the situation is the one-sided "better safe than sorry" commitment. For then attending to logic only leads to a conclusion your intuition says is wrong. And for every one of us, it is almost always easier to distrust the logic of an argument than to distrust our own intuition. Hence the problem, from the risk matrix point of view, is to find some way in which a rival to one-sided perceptions (when, in fact, a rival is logically relevant) can be built into the process. What is needed are process changes that put awareness of costs of precautions routinely alongside awareness of danger. A person might still want more caution than the experts think is needed. So, fungibility does not automatically determine a result, nor should it. But it opens the choice to reasoned discussion.

In particular cases, fungibility might be prompted by circumstance or by a sufficiently intense social effort. But the former cases will be rare, and the latter can occur (also rarely) only when sufficiently powerful or influential actors are motivated to make an extraordinary effort, the case of genetic engineering being one examined here (6.9). But it is sensible to want to make it easier for good arguments and good evidence to influence outcomes even if the circumstances are not unusually favorable.

7.4

Fungibility is missed when individuals somehow are insensitive to one side of a risk issue. So the natural aim for reform proposals is to put opportunities forgone on-screen (if it is opportunities that are somehow being ignored, and only danger attended to). As has now been discussed at length, a double impediment makes that ordinarily difficult. One difficulty is that the missed side of the balance—the costs of precaution that would have to be accepted—is outside a person's direct experience, where it is hard to see.[2] And the complication I have been stressing is that if fairness

2. As I have repeatedly mentioned, the costs (or benefits forgone) of *not* imposing the regulation are also likely to be outside a person's experience. But just the cases that become the focus of controversies are those where a default "better safe than sorry" propensity is prompted from social "everyone knows" knowledge that anchors perception of the substance at issue in danger and guilt.

intuitions have fallen in line with "better safe than sorry," as indeed will usually happen, then an appeal to balancing costs and benefits, instead of seeming obviously sensible, will easily seem intrinsically unfair and hence perverse.

Consequently, a critical element for reform proposals will be whether they encourage noticing fungibility, but in a way that does not defeat itself by provoking a moral response that makes talk of efficiency seem irrelevant or perverse. To restate that important point: it will be especially hard to promote fungibility if that effort is vulnerable to attack as just an attempt to rationalize why victims be allowed to suffer so that those who are not victims can prosper even more than they already do. And of course that turns on what is cognitively realistic, which is not the same thing as what is logical.

Politically sensitive actors will (of course) be slow to align themselves with a position that might be seen as, or might be effectively claimed by their rivals to be, on the side of exploiters and indifferent to victims. A sensible political actor cares a lot about how things look to his audience. Political actors sometimes respond as if they scarcely care at all about what is really at issue, only about what the issue looks like to the public. For in fact, failing to attend to such matters is a formula for elimination from political life. So a challenge for the "do no harm" proposal that I will take up in chapter 8 is how to build intrinsic concern for doing no harm so firmly into the process that politically vulnerable actors do not have to take responsibility for arguments that will sometimes be attacked as just a roundabout way to harm victims.

But that same problem arises in a more generic fashion in the context of public mediation of environmental disputes. No one with a political stake wants to defend the unpopular side of an issue if that can be avoided. And even among individuals who appear to have a major stake, each has other problems on his plate and a limited ability to control the social choice anyway. It then easily turns out that no one sees working toward the most socially reasonable outcome as her main concern. So if it is politically or socially awkward to raise questions about the costs of precautions (because that is so easily seen as signaling indifference to victims), then that side of things may go largely ignored, which once again requires attention to process changes that might enhance the visibility of fungibility. So we will be concerned (in chap. 8) with a process change that puts information on the table that might enhance fungibility, and (in chap. 9) with putting an actor motivated to push attention to fungibility into the public participation process.

7.5

We need to attend to just how "better safe than sorry" intuitions work their way through the regulatory process. If fungibility is absent, by definition a person is not seeing anything to balance. Costs of accepting the risk (even if ephemeral) are on-screen, but benefits (even if substantial) are not, in the way already discussed in detail. Or, as I have often mentioned, benefits may be on-screen and costs off-screen, but missed fungibility of the first sort is what is salient here.

Difficulties then enter in at least two forms beyond those so far discussed. First, "better safe than sorry" intuitions enter deep into the process of assessing the risk. That can make the process cautious in the extreme, since the eventual risk assessment can then reflect compounding of a considerable series of cautious judgments, each motivated by "better safe than sorry." (As will be seen later, this, in turn, interacts with a second factor, which turns on twoness responses to the results generated by that compounding of "better safe than sorry" responses.) But sensitivity to the competing concerns of dangers averted and opportunity lost will also by influenced by a person's social situation and incentives. A prudent person would not feel reassured that a nuclear waste proposal was safe solely on the judgment, however sincere and technically expert, of people employed by the nuclear power industry. Committed supporters of nuclear power may be able to make an argument that is convincing to sophisticated experts not so committed. But the lay person is not convinced by the judgment of industry experts alone. Even if only unconsciously, they are too committed to one side, as you and I also are vulnerable to bias on issues in which we have a stake or we have social relations or even only a habitual disposition that makes it much easier to see one side of an issue than the alternative.

But the other side of this coin is that a person personally feeling one-sidedly "better safe than sorry" is not going to be easily alerted to the *costs* of one-sided "better safe than sorry" commitments. And indeed even a person ordinarily immune to that, but in a social context (such as serving on a committee reporting to principals who are focused only on risk) where "better safe than sorry" is the taken-for-granted sense of the situation, is likely to *become* insensitive to those costs. In general, working on a risk assessment need not engage at all any sense of the costs of precautions, and often—indeed, ordinarily, unless the work is done under the intrinsically suspect auspices of someone who stands to gain if tighter regulation is not needed—that work is done in a setting where concerns about costs do not easily arise and cannot easily be sustained if they do arise. Worrying about costs of possibly excessive caution is someone else's department.

And most expert participants in the *details* of a risk assessment are not expert at a level where comparison across risks, or comparison of costs of accepting and costs of avoiding a risk, are yet an issue. They are not likely to be vulnerable to the severe twoness effects that influence lay intuition. But they are also not likely to exhibit concern that one-sided "better safe than sorry" estimates might be misunderstood—for example, taken as best estimates, not as very cautious estimates—at later stages of the process. So we easily get the compounding of "better safe" judgments already mentioned.

The consequence is then that when a standard is described as involving a risk of (say) 1 in 100,000 of a fatality, a more realistic characterization of the standard will often be that there is a risk of 1 in 1,000 that any individual will be exposed to a risk, which on highly pessimistic assumptions (each assumption "better safe than sorry") might be 1 in 100,000—which means we are really talking about a cautiously estimated risk of 1 in 100,000,000. I want to soon sketch out some detail of how that happens, since noticing how it happens is important for seeing the point of the "do no harm" proposal I will take up in chapter 8.

And a second class of difficulty then severely aggravates matters. Almost no one but an economist feels comfortable talking about what a human life is worth, hence what it would be reasonable to pay to avoid a situation that involves a risk to human life. Since we are mortal, there is nothing we can do that would literally save (or lose) a life. We are always actually talking about a probability of extending (or avoidably shortening) life by some amount. But the usual shorthand is to talk about the value of a life. That presumably aggravates, though it does not plausibly create, the difficulty of talking about what extending or shortening a life is worth.

Economists like to point out, and correctly point out, that every individual and every society in fact makes value-of-life choices all the time. And the implied valuation revealed by individual and social choices of human life is often very low compared to what is implicitly in effect even for environmental regulations that are *not* controversial. But we can observe that the valuations, to be acceptable, have to be kept implicit. A visitor from Mars, watching what we do, could only conclude that we do not mind trading off dollars for lives, we just do not want to be pushed into noticing it. Or, more generously: the world is such that there is often nothing we can do that does not implicitly involve trading off lives for dollars, so of course we do what we cannot avoid doing. But we hate to be pushed into noticing it.

That reluctance then interacts with the propensity to twoness in judgment (4.7). Calculations of environmental standards are routinely done in the exceedingly cautious way just sketched. But that does not engender caution about interpreting standards. Rather, the polarizing ("twoness") tendency yields either/or perceptions: the risk level chosen as the standard is seen as a bright line, dividing safe from unsafe. On either side of the line, there is no occasion to think about how much we are willing to spend on a human life. Below the line, no danger is seen, hence nothing need be spent. Above the line, the standard simply must be met regardless of cost or circumstance (e.g, that the risk is only that of a transient exposure above a standard conservatively set to cover lifetime exposure).

What weakens and sometimes reverses this propensity to extreme caution, as various examples in chapter 6 illustrated, is not at all likely to be an appeal to logic, though that may come to be crucial at the next stage. Rather, it will be circumstances that prompt fungibility by tying some hitherto neglected aspect of a situation to direct experience. But calculating what a life is worth is *still* avoided. Things simply look different, and what was obviously dangerous comes to seem obviously trivial.

7.6

Now consider some detail of how what seems a bright line marking the transition from safety to danger is constructed. Substances suspected as human carcinogens are first calibrated to find the highest exposure that will not kill the test animals before they can live long enough to develop cancers. The animals (usually mice or rats) are then tested at some substantial fraction (most commonly, 1/2) of that toxic dose. The point estimate to start a calculation of cancer incidence in humans is then derived from the frequency of tumors (not necessarily malignant) in the most vulnerable organ of the most sensitive species tested.

This criterion is severely employed. If a substance increases tumors in one organ but decreases tumors in another, the damaging effect is used and the protective effects are ignored. At near toxic doses, we are not very likely to see benign effects. But at less intense doses—which also means at doses not so vastly remote from the trace contamination that is the usual concern for regulators—benign as well as adverse effects are in fact very common. But benign effects are routinely ignored, not by neglect, but as a matter of chosen procedure. "Better safe than sorry."

If that worst effect occurs in mice but not in rats, then the extrapolation is from mice to humans. If the effect is on rats but not on mice, then rats will be the basis for extrapolating to humans. About 50% of substances, with about equal likelihood for natural as against manufactured substances, turn out to be carcinogenic for some organ of some species at some sufficiently high dose. But as closely related as they are, about 40% of the substances that are carcinogens for either mice or rats are not so for both. Since mice and rats are much more closely related than rodents and humans, and since various other "worst case" assumptions have been made or conditions imposed (such as ignoring benign effects), it is a strong assumption indeed to suppose that the most severe effect on the most severely affected animal translates into a realistic estimate of effects on humans.

Particular numerical estimates are calculated by extrapolating the risk to the most susceptible rodent at near lethal doses down by some huge factor to estimate effects on people at doses of concern to regulators. The extrapolation process is complicated, but a critical feature routinely invoked is that damage is effectively taken to be linear. Combined with the consideration only of adverse effects, that builds into the assessment a commitment to treating any dose whatever, however small, as creating a calculable risk of cancer.

But (a counter argues) would that caution not be balanced by cases where humans, in fact, are susceptible but mice or rats are not? Yet there is no reason to suppose that things will balance out. Overregulating some risks (those which produce effects on rodents) does not somehow offset underregulating others, any more than driving 30 miles per hour under the speed limit half the time balances out the risks of driving 30 miles per hour over the speed limit the rest of the time.

7.7

The "healthy worker" effect is a concern in epidemiological studies which prompts caution in extrapolating from effects on workers to effects on the general population. Workers, on the whole, are healthier than the population in general, since workers do not include the elderly, the sick, the very young, and various other high risk groups. So estimates drawn from exposure of workers will easily underestimate the effect of the same exposures on a general population. But extrapolating from animal tests at near-toxic doses prompts a reverse concern with what might be called the "sick rat" problem, which is especially likely to overestimate dangers given the usual linearity assumption.

The linearity assumption is not guaranteed to be a worst case. But the predominant view among experts in the etiology of cancer is certainly that it will ordinarily exaggerate the effects and easily do so by a large amount. There are multiple issues here, of which the following seem to be the main points.

There are, first, various aspect of the "sick rat" problem. Near toxic test doses kill cells, prompting abnormally fast cell growth to offset that, which is itself a major risk factor for cancer. Another such issue is that DNA damage is going on all the time, mainly from unavoidable natural insults. None of us could live long except that the body also has elaborate repair mechanisms, which are at work all the time. But the severe doses involved in testing for carcinogens can overwhelm those normal processes. Third, normal processes for metabolizing, storing, and excreting the substance are also likely to be overwhelmed by huge doses, so that, for example, some fraction of a damaging metabolite may end up being stored in a part of the body that ordinarily (i.e., without this overwhelming challenge) would never have to face that insult.

The consequence of this is a point already mentioned, which will be important for the general argument to come: at low doses (i.e., at doses well short of what simply overwhelms normal processes) it is commonplace for substances to produce *mixed* effects: helping a bit on one side, harming a bit on the other. This is true even for substances that we think of as essential (so if the doctor fears you are not getting enough, he will prescribe something to get it to you.) But since the risk calculations ordinarily consider only the worst effect, and indeed at massive doses, we are not likely to be able to see any benign effects. On almost any dimension we might study the dose is now large enough to make the substance a poison. But it is common—not an oddity occasionally noticed, but common—that animals at intermediate doses do better than control animals at zero dose. A mix of results is commonly found so that adverse effects are at least partly and quite often more than wholly offset by benign effects.[3]

The overall effect of all compounded cautions (such as ignoring benign effects) suggests that the calculated effect can easily be a large multiple of the actual effect, and may even have the wrong sign. But even if that were so in every case, one could not prove it. So although illogical once the significance of ignoring benign effects is noticed, the usual sentiment is that if you want to be surely "better safe than sorry," you just cannot be too cautious. Although I will not pursue details here, methods similar in spirit are used to calculate effects of noncarcinogens. But symptoms of immediate toxicity are usually at issue here, not the long dormancy period of cancer, and there are more likely to be extensive data available on dose/response relations. So opportunities for compounding pessimistic assumptions are not so nearly unbounded. On the other hand, for noncarcinogens, it is the usual practice to include at least two, and occasionally three, multiples of ten as a safety factor.

7.8

The assessment process just sketched does not exhaust the conservatism built into regulatory standards. A further set of questions comes up when a standard is applied in practice. It is easy to imagine situations in which a high concentration of a potent toxin would pose no unreasonable risk: for example, it is in a bottle, properly marked, in a laboratory. Similarly, what might be considered no risk—say a bottle of aspirin—could be fatal under various circumstances: a person under suicide watch in a mental ward has access to it, or a small child has access to it. So necessary questions in setting a standard for some substance (or some activity) will turn on how cautious to be about considering the most unfavorable conditions that need to be protected against. The "better safe than

3. There has been extended debate over these long-noted effects. See the exchange of correspondence in *Science* (1988).

sorry" pressures that yield risk estimates compounded from many pessimistic assumptions have parallel effects in judging the impact of the risk, given a level of concentration.

A real but not atypical example is that, for nuclear plants, radiation risk is calculated (using the compounded pessimism already sketched) for a person who is born at the plant boundary, and spends the next 70 years (no time off for vacations, school, or work) at that location. Naturally, no one supposes that any such person will exist. But "better safe than sorry."

A usual response here is that if harm is linear, then it makes no difference whether one person is exposed to the risk for a lifetime, or many different people a total of 70 years. But in most actual cases of this sort, either harm is not very plausibly linear, or there is no such set of people (e.g., persons dwelling for a cumulative lifetime at the boundary line of a nuclear waste site or directly atop an abandoned nuclear reactor). In any case, the rationale for concern about a risk that is one in a million (what is commonly taken as sufficient for regulatory protection) is that many people would be exposed to this risk, not that an entire population, in the aggregate, is exposed to a single risk of one in a million of one additional cancer.[4]

When exposures are calculated that turn on gradual seepage from a storage site, pessimistic assumptions are again compounded, attempting to reach a result that with high probability could not be exceeded. In fact such calculations have sometimes proved to be not pessimistic enough. The most notorious case is probably radiation leakage from the West Valley, New York, storage site. On the other hand, the seriousness of the leakage as a health threat in that highly publicized case is open to severe question, as implied by the title of a skeptical appraisal: "Safer than Sleeping with Your Spouse" (Matuszek 1988). But, as has now often been mentioned, twoness effects enter, and any leakage whatever beyond what is explicitly allowed under a standard be-

4. For a detailed and balanced discussion, see Roderick (1992). For a blunter and more recent view, see Abelson's editorial in *Science* (9 September 1994).

comes a transgression of the bright line between safety and danger.

7.9

What compounding "better safe than sorry" assumptions comes to is that estimated effects are not effects to be expected, but effects calculated in such a way that it is unlikely (though of course it cannot ever be proven absolutely impossible) that any substance treated in this way could be more hazardous than the regulatory model predicts. It is as if, when you packed for a trip, you prepared for every contingency you could imagine and you worked hard at imagining contingencies. It would not be surprising if you then later found you had packed more than you needed. But all this is just what will make sense to a person for whom fungibility is absent. You do not see the costs of caution, only the bleakest costs that might conceivably be incurred without that much caution. Effectively, then, what happens is that caution compounds caution (there is really no limit to how cautious you could possibly be, there is always some further contingency you might pack for) until things become so stark that benefits of not being *that* cautious finally are forced on-screen. As proposed standards get tighter, the number of people who would be adversely affected goes up, and so does the intensity of the costs of precautions for those who have to pay those costs. Sooner or later, this yields countervailing effects: but very often far later than is easy to make sense of in terms of how people ordinarily want resources used.

Once fungibility enters (if it does), the compounded precautions can come to seem entirely unreasonable. That can be seen in the way ordinary people talk, and in the way journalist and politicians talk, where arguments that, in one context, are treated as just evasions (when the discussion is about something conspicuously failing to prompt fungibility, such as dioxin or a nuclear waste site) in another context (for example in the overwhelming public reaction to proposals to ban saccharin) are seen as mere common sense. Then the hypercautious extrapolations from the equivalent of 800-cans-of-diet-soda-per-day doses to rats to part-per-billion doses to humans are treated as weird arguments that only zealots or bureaucrats could take as sensible. We see, yet again, the twoness effect: what had been taken as a bright line between safety and danger now looks like a risk level too trivial to be taken seriously.

A further problem arises if "better safe than sorry" propensities routinely push toward very tight standards. Those who are concerned about overregulation devote effort to keeping their item off the regulatory agenda entirely. Even without any active lobbying to fend off regulation, there are far more candidates for regulation than regulators have time and resources to handle—a situation aggravated by the propensity to overregulate if regulation is undertaken at all, which maximizing the incentive of those subject to such regulations to fight hard against them. Occupational Safety and Health Administration (OSHA) regulations under the Reagan administration were not noticeably less severe than under Carter administration (see Mendeloff 1988). But there was a general slowdown in the rate at which new regulations were moved through the process. So, in Mendeloff's phrase, overregulation (of what gets on the agenda) yields underregulation (of many things consequently left off the agenda).

7.10

If enough people feel worried about some risk, however remote and cautiously calculated, then it makes sense to say that the government ought to respond to that. How to respond is less clear. But no one is likely to explicitly argue against a claim that we will generally do a better job of getting what we want if we see what choices we are actually making. Confronted with the chips problem of chapter 3, a reader will usually come to see that the logical answer indeed must be 2/3. But even after that takes hold, it can still take a long time before the same person feels comfortable with what logically she now believes: for some time, the probability continues to *feel* like it must be 1/2. On far more complicated judgments on environmental risk, beset by far greater uncertainties, tied to wider commitments, and easily provoking strong emotions, it is very much harder to escape a one-sided "better safe than sorry" intuition. Once that is in place it will not be displaced merely by a logically strong argument about fungibility. In situations well outside direct experience, it is remarkably easy to be cognitively blind to information that, on our own view, is logically relevant to our judgment or to arguments that ordinarily we see as important. We are not in principle against using information or logic to inform judgment, just in practice blind to it in such situations. But that in principle we are open to information and reasoned argument, though in practice sometimes blind to it suggests we might look for remedial proposals that somehow manage to appeal to that openness in principle to help overcome the blindness in practice.

Consider, therefore, a deliberately modest reform, in which the costs of precautions pushed to the fore are not the set of all logically relevant costs, but only some subset that is cognitively salient. In particular, they are a subset selected because for one reason or another *these* costs are more easily seen as the kind that ought to be relevant to sensible choices about risk. Their status is so clear that their legitimacy holds even in a context strongly marked by concern for the plight of victims.

Rules that look reasonable can yield results that look wrong. In our context, agreeing ex ante to follow a certain procedure would not preempt one-sided "better safe than sorry" responses in a particular case at hand. But now those "better safe than sorry" intuitions conflict with other intuitions (about fair procedure) that say these "better safe than sorry" intuitions, hard as they may be to doubt at this moment, might be mistaken.

Proposals that assessment of risk should include assessment of the costs of averting the risk are commonplace (7.2). The novel aspect I will seek to develop here amounts to insertion of an intermediate stage. This stage would be intermediate between "better safe than sorry" consideration only of whatever risk is the focus of public concern and the full analysis of costs and benefits that makes sense to economists. That intermediate stage involves not the balancing of all costs and benefits, but only the subset that can appeal to rather than repel lay intuitions about what fits with a fair rule about how in general, to proceed. Then fungibility of a sort is built into the process: the rules look right, even in a particular case where the result feels uncomfortable.

"Do No Harm"

8

A first rule for regulators, as Hippocrates famously specified for physicians, might be "do no harm." The risk matrix argument suggests a particular sort of "do no harm" rule. The rule itself is only that *if a regulation is intended to protect public health, then there ought to be reasonable confidence that the effect of the regulation in fact would be to save lives, not cost lives.* But an interpretation of that rule requires some discussion. Since it derives from the risk matrix, not from economics, the rule as proposed here is not intended to cover anything like the full range of consequences that would be considered for economic analysis of what also could be labeled "do no harm." For example, "trickle-down" effects (8.8), which have gotten a good of attention from economists and some attention from the courts as a component of what economists call "risk-risk" analysis, are unlikely candidates for the "do no harm" assessment proposed here. But the general notion of "do no harm" is so close to immediate common sense that it has been discussed in one way or another many times. Graham and Weiner (1995) provide a stimulating set of papers on the theme. The novelty of the notion as discussed here turns on its details, tied as they are to the cognitive analysis of the risk matrix.[1]

1. After this chapter was in draft, I came across a similar proposal by Stephen Breyer and Thomas Ehrlich in their minority comments on the National Academy of Sciences saccharin report (SC 19–22 of pt. 2 of the Food Safety

Like its sister proposals, the proposal here turns on the point that actions have unintended as well as intended consequences. For a well-marked risk, those unintended effects (side effects) are usually small compared to the main risk. But for the subtle risks characteristic of the expert/lay conflicts (5.10), there is no reason to simply suppose that the effects on by-product risks will be smaller than the risk the precautions are intended to correct. There will be a real chance, not just a conceptual possibility, that the side effects will turn out to have more prospect of hurting people than the intended effect has of protecting them.

Since that is the only issue for "do no harm" as proposed here, the proposal provides only a minimal, not a sufficient assessment of the pros and cons of a proposed regulation. It ignores dollar costs per se, and would exclude even the (imputed) health impacts of income loss that might be charged to regulation. But the proposal does aim to give the "do no harm" question in at least that modest form legal and moral standing. And for reasons I will try to spell out, that is not a small matter at all.

8.2

Not even the most health-conscious person or society treats any increase in risk to health, however small, as outweighing any cost, however large. But various major pieces of environmental legislation, on their face, have implied that by ruling out any consideration of costs. In practice, as costs grow, an exclusive focus on health yields to countervailing pressures (6.3). But if spread thinly, even very large dollar costs can remain politically imperceptible. What logically ought to be far more than sufficient to prompt awareness of fungibility on a risk issue will sometimes be cognitively ineffective. That has by now been discussed in detail. Logically, fungibility is always present: there are no free lunches. But cognitively, fungibility is easily missed. "Do no harm," as

Study, March 1979). No doubt similar earlier proposals could be found. The novelty here is entirely in the cognitive viewpoint that guides the detailed discussion. On risk-risk analysis, see the series of articles in the special issue of the *Journal of Risk and Uncertainty* (January 1994), edited by W. K. Viscusi. See also Lave (1981).

suggested by the concluding discussion of chapter 7, seeks to build a measure, albeit only a partial measure, of fungibility deeply into the *process* of risk assessment and risk management in a way that appeals to everyday intuitions about fairness.

One might suppose that a "do no harm" rule in some form was already built into risk regulation. On its face, asking that regulations intended to save life should, at a minimum, be more likely to save lives than destroy lives sounds like what any sensible regulation must be intended to do. But if such language were written into legislation—in the widest form, as a Congressional resolution about how legislation governing risk regulation was intended to be construed—the effects would be large. Agencies and courts would often have difficulty showing reasonable confidence that the effect of their decisions in fact would be to save lives, not cost lives. A legally and socially manageable version of "do no harm" would have to be carefully crafted to control wholesale challenges of existing regulations and decisions.[2]

Substantively, repeating that point, the proposal here is deliberately modest compared with a version of "do no harm" (risk-risk analysis, mentioned earlier) designed to push as many costs as possible on-screen in contexts where the regulatory process explicitly or effectively rules out a full cost-benefit analysis. In a context where there is wide regulatory use of cost-benefit calculations (a possibility as this goes to press, in the wake of the 1994 Congressional elections) risk-risk analysis would be redundant. But the proposal here is designed to offset a cognitive, not a legal or administrative, difficulty. Consequently, its role would not disappear in such a regime. For its aim is to complement not substitute for fuller analysis.

8.3

Consider "do no harm" first in the context of the more familiar reform proposals reviewed in 7.3. The many efforts to set out

2. For a salient example: it is hard to imagine how any risk attributed to pouring Hanford well water onto the ground adjacent to the well from which it had just been removed could exceed the risk (small too be sure, but not zero) of someone being injured in the process of trucking the water around the countryside (6.4).

risk comparisons from Starr's pioneering article in 1969 through Breyer's 1993 book and still more recent work reflect a sense that risk regulation would work better if the process had somehow built into it a standardized yardstick for risk, analogous to the constant dollars of economic comparisons. The "do no harm" proposal is in a certain way a functional equivalent of a risk comparison. The comparison, however, is not with other risks (each risk taken independent of context) but with other risks involved in the very situation at hand.

"Do no harm" also would contribute to a marginal returns perspective, though it is not *directly* about concern with diminishing marginal returns. It focuses attention on what might reasonably be expected in terms of net health benefits of a marginal change in regulations, but in a way that raises no issue of lives against dollars (7.2). In the context of "do no harm", doubts about prudent design of regulations would not take a form equivalent to putting a price on what a life is worth. Often, a careful assessment would lead to a range of possibilities that includes saving lives by regulating less.

And sometimes "do no harm" would be functionally equivalent to a *de minimis* standard, since a criterion reasonably considered in deciding whether a risk is *de minimis* would be if the risk at issue is so small that we cannot even get reasonable confidence that further controls (if imposed) would save more lives than they would destroy. The smaller the risk that is being diminished (the more nearly *de minimis* the problem), the more likely it is that in fact no clear benefit shows up in a "do no harm" assessment.

To illustrate how "do no harm" might work, an old episode— the abortive FDA plan to ban saccharin—provides an easy case already mentioned (7.10) . The current (unresolved as this is written) dispute over trace dioxin contamination provides a harder and more nearly typical case. Saccharin provides an easy case and dioxin a hard case for the usual reason: it is easy to link the costs of banning saccharin to ordinary experience and very hard to do that for the costs of yet tighter controls on dioxin. So fungibility comes easily in one case but not in the other, with consequences we could expect.

8.4

Clearly, if we want reasonable confidence that a health regulation would have overall good effects, then the various ways in which the regulation might affect health have to be estimated in a manner that does not stack the deck in favor of the regulation. So for saccharin, the risk of cancer would have to be weighed against possible benefits of saccharin, weight loss from avoiding the calories in sugar being the most obvious. And that balance, obviously, could not use a dual standard that treated some risks (those from using saccharin) as serious unless proven innocent but others (those from not using saccharin) as trivial unless proven guilty.

A counterargument is that the public dreads some bad things more than others (say death from cancer as against death from another disease), so that it is appropriate that special weight (under the Delaney Clause, even unlimited weight) be given to carcinogens. But we have already had occasion to notice that on concrete cases where fungibility is not lost, the public reveals much more balanced views.[3] That can in fact be seen on the saccharin matter here and on various cases (e.g., aflatoxin) discussed in chapter 6. In any event, even "better safe than sorry" responses do not always weigh cancer as a risk more consequential than other risks. And special concern about cancer could not account for cases (dioxin will provide an example), where some cancer risks are simply ignored when attention is focused on other cancer risks in the very same context.

For saccharin, there is in fact great uncertainty about how far

---

3. A component of work defining "do no harm" might include opinion surveys to assess how far public sentiment in fact would favor policies that *as a matter of principle* harm more members of the public then they help, provided that the help is on health issues of special concern (say, cancers or particular forms of cancer). I would be skeptical that large disparities would appear that are robust across alternative ways of framing the questions. If that could be shown, however, it would be in the spirit of the proposal here to seek to accommodate such preferences. On the other hand, it would not at all be in that spirit to take concern exhibited in some particular case as a demonstration of what makes sense in that particular case. See the concluding remarks of this chapter.

use of reduced calorie sweeteners actually reduces weight. But there is also great uncertainty about how far saccharin increases the chance of cancer. The only carcinogenic effect clearly seen was on bladder cancer in male rats subjected to a lifetime diet of at least 5% saccharin after their mothers also had been fed a lifetime diet of at least 5% saccharin. A human being (*and* his mother) would have to drink 800 cans of diet soda a day to obtain a comparable dose (OTA 1977, 22).

On the other hand, clearly there is some offsetting effect to using saccharin that leads to more calorie consumption in other parts of the diet or to some shift in metabolism. Otherwise a person not currently using nonsugar sweeteners could easily lose a dozen pounds over the coming year merely by switching, which many a reader who would prefer to weigh less has discovered is wrong. But the health effects of extra weight are very marked. So there could be very substantial aggregate effects integrated across the large population involved (on morale and fitness as well as on lowered probability of heart attacks, stroke, and the many other weight-related diseases and disabilities), though the average effect per individual might remain too small to show up in epidemiological studies. Even if only a modest fraction of a pound (not a dozen pounds) were lost, the extrapolated favorable effects on life expectancy could be much larger than the carcinogenic effects. And the favorable effects extrapolated from overt excess weight in people to a few ounces of excess weight in people would surely be no less certain than bladder cancer effects extrapolated to people from massive (800 cans a day equivalent) lifetime doses to two generations of rats.

In any case, we would not get a meaningful comparison unless some care had been taken to construct the estimates in comparable ways. The effects of treating dangers by different standards need not be at all subtle. Suppose that, as in fact the matter was treated not only by the FDA but also by the report of a National Academy of Sciences panel on the issue, even weak indications (say, distant extrapolations from very high doses in rats, or markedly flawed epidemiological studies) that saccharin is carcinogenic was taken as evidence that cannot be ignored ("better safe than sorry"), and evidence that saccharin helps control weight was seen as properly ignored unless unambiguously strong.[4] Then the benefits of saccharin could outweigh the losses a thousandfold and a decision be made to ban it anyway. One side is guilty unless proven innocent; the other is innocent unless proven guilty. Uncertainty adverse to saccharin is resolved by focusing on the possibility that the threat might be real, and uncertainty favoring saccharin is resolved by focusing on the possibility that the benefits may be only illusory. The cumulative effects of such treatment could easily be very large.

So in the saccharin case, the possibility arose of a huge discrepancy between the FDA judgment, guided by the Delaney Clause, as against a widespread public perception of a need to balance the risks. The effects of saccharin use are sufficiently subtle that clear evidence of what effect it has on weight remains elusive. But the general problem of being overweight is not a subtle risk. For all too many of us, it is a matter easily brought to attention by looking into a mirror. Of course, a critic could respond by saying that, realistically, what is at issue is an average effect on weight (spread across all saccharin users) that might be only some small fraction of a pound. But cognitively, the anchor for judging that risk is a very familiar and often directly experienced problem of evident overweight, as cognitively the anchor for judging a one-in-a-million cancer risk is likely to be a person you know who definitely has cancer. Extrapolating from an obviously overweight person to an effect of only a few ounces is hardly low confidence relative to the rats-to-people extrapolation on which the cancer risk is based. More generally, the whole set of

---

4. The OTA study (1977) which preceded the Academy study, and which (like it) was specifically directed to consider benefits as well harm from saccharin, suggested that the availability of saccharin might in fact increase consumption of sugar (p. 34). The reasoning was that the more sweetening that was available, the more people would come to like sweets, and so they might eat more sugary sweets than if there had never been a sugar substitute available. No comparable skepticism is offered about the confidence that could be put in extrapolations from dosing two generations of rats with the equivalent of 800 cans of diet soda per day.

usual arguments about why a person might reasonably worry about a very subtle "risk of a risk"[5] of cancer reveals nothing that cannot be paralleled by an at least equally good argument for paying similar attention to comparably subtle risks of heart attacks or stroke, among other health risks clearly tied to weight.

So the argument is not that subtle risks should always be ignored, only that subtle risks of one sort should be assessed parallel to other subtle risks affected by the very same choice. We ought not to see intense attention to one risk and casual dismissal of others. But if we do take comparably seriously the array of risks in a situation, we introduce a measure of fungibility. We have to consider how far it makes sense in this case to act in a way that increases some risks in order to decrease others.

In the saccharin case we can observe what the earlier argument would lead us to expect. Common experience made fungibility apparent. That awareness of fungibility prompted vigorous public opposition to banning saccharin, since it was then easy to judge that the ban could easily mean trading more weight for a very tiny prospect of avoiding cancer. And the earlier argument (5.9) then suggests that framing and fairness intuitions would come to support that intuition.

Indeed, public and political comments on the issue leave no doubt that, in this case, most people came to see the status quo as including the artificial additive, so that eliminating a possible cancer risk would be a gain from the status quo and eliminating an aid to controlling weight would be a loss. But there was nothing *logically* inevitable about that framing: absent a perception of fungibility, saccharin could easily be seen as an artificial alteration of the natural situation. And absent fungibility, it would be easy to see saccharin as a risky (for consumers) way for profiteers in the soft drink industry to seduce victims into spending more on their product. With fungibility, the same situation is more easily

5. The phrase is from Breyer and Ehrlich's comments (see n. 1).

seen as a matter of making ordinary people the victims of zealous bureaucrats.[6]

So in the way already discussed (5.9), all three Fs (fungibility, fairness, and framing) in fact did come to reinforce each other in conflict with the FDA proposal to ban, and with the leading role played by fungibility, as should be usually expected. Consequently, in the case of saccharin a one-sided "better safe than sorry" assessment did not prevail. Because so many individuals know from direct, firsthand experience about weight problems and care about that, fungibility could not be missed on this issue. Saccharin survived the proposal to ban despite official assessments that stacked the deck, because public protest overrode the usual result.

In Aspen, with fungibility again stark (6.3), attention was paid (by the community, but not by the EPA) to the risks involved in heavy trucking over the twisting road to the town, to doubts about the EPA's very cautious calculations of the risk from lead, and indeed to the residents' preference for living in Aspen. As already discussed (6.3), framing and fairness intuitions again demonstrated their propensity to fall in line with whatever emerges from fungibility. For lead in Aspen as for saccharin, government agencies eventually got into trouble, not because the choices made were at variance with what is ordinarily demanded of these agencies, but only because circumstances strongly prompted fungibility in a substantial enough segment of the public: for the case of saccharin, spread across the whole country, for mine tailings at Aspen, among the only people with an immediate stake.

It is instructive to compare what happened in Aspen, Colorado, and Triumph, Idaho, with what happened in Times Beach,

---

6. This looks like a splendid case for voluntariness as a cause of risk perceptions (here consumption is voluntary, and risk is indeed perceived as low). That is clear. Yet even in this (apparently) clear case, for people seeing things from a "better safe than sorry" perspective, perceptions of voluntariness are not so clear. The NAS report scarcely notices the voluntary aspect. Public health advocacy groups that favored the ban argued that saccharin was *not* really voluntarily: people were frightened or misled into using it, children were trained into the habit of using it, etc.

Missouri. Times Beach was effectively destroyed incidental to an EPA order to remediate roads that had been contaminated with dioxin. The situation of residents of Aspen and Triumph was helped by the fact that it is easier to demand evidence of symptoms of harm for metals contamination than for cancer with its long latencies. But citizens were probably also better able to prevail because they had wealthy and influential employers and neighbors. So it was harder for paternalism to prevail over local sentiment.

On the other hand, it is important to see that the EPA's paternalism was not simple big brotherism. The most important influence on the EPA, as Justice Breyer's "vicious circle" argument urges, came from acting in the way that bureaucrats have learned they usually *must* act to stay out of political trouble. But does it make social sense that it should take extraordinary circumstances for regulators to behave reasonably? Recall here that when children were tested in Aspen and Triumph they did not in fact show elevated blood levels, and local hospitals had seen no unusual prevalence of toxic symptoms among people with homes built on the mine tailings.

8.5

The Delaney Clause requires that the FDA ban any substance added to food that is a carcinogen. and whether a substance is a carcinogen is judged in "better safe than sorry" fashion (7.5). The FDA has no explicit authority to consider how strong or weak the carcinogenic effects might be, nor whether there may be offsetting benign effects. In practice, that is hedged in various ways. Sufficiently gross situations will prompt tactics that avoid bizarre outcomes. On the face of things, chlorination of water violates the Delaney Clause. But there is no serious alternative. Hence on this matter, the FDA has taken a narrow reading: adding chlorine to water inevitably creates some chloroform, a known carcinogen at high doses in laboratory animals, but chlorine itself is not a known carcinogen. So in a sense, adding chlorine is not adding a carcinogen. On this reasoning the FDA could avoid the chaos that would arise if it claimed it was obligated to forbid chlorination. But the same reasoning, if a proper interpretation of Delaney, would put in question many other FDA choices since the FDA would not in general ignore a suspected carcinogen if it is only a derivative of an additive, not directly the additive.

In more typical situations, Breyer's "vicious circle" prevails. There is pressure on regulators to perform in a one-sided way, since, when "better safe" prevails, a balanced view gets them into trouble with the press and the public (hence with legislators) even in cases where a balanced assessment in fact is allowed under current law, which it often is not. It is only when the situation is such that "better safe than sorry" yields a policy so grossly against common sense that a proposed ban would cause political turmoil (as would a proposal to ban chlorination of water) that something like "do no harm" comes clearly into play.

8.6

Saccharin is an easy case for "do no harm," in the sense that not only logically but (more important) cognitively, it is so clearly a case where some fungibility judgment is needed. Consuming fewer calories is, to say the least, a reasonable strategy for losing weight. Saccharin is thoroughly associated with this benign aim, and familiar (so particularly for a user of saccharin, the framing of the status quo readily includes saccharin), and it is consumed voluntarily by people pursuing the benign aim of avoiding excess weight. Its emotional coloring, consequently, is neither danger-ous nor guilty. Using saccharin is easily framed as a prudent, not reckless, continuation of the status quo. So in this case the need for some "do no harm" assessment was cognitively easy to accept: in fact, the public sense of things made it impossible to deny.

Dioxin comes at the other extreme of the spectrum. It starts from associations strongly colored by both danger and guilt (pesti-cides and the Vietnam War). And no one (not even people with a favorable view of pesticides or the Vietnam War) ever wanted dioxin: it is an unwanted by-product of other processes. Both logically and cognitively, consequently, it is a hard case as conspic-uously as saccharin is an easy case.

On the other hand, it is also a case which provides a striking example of Breyer's "last 10%" concern. Because obvious sources of dioxin have long been subject to strict controls, what is left are numerous secondary sources, each contributing a minuscule amount to the environment. Total emissions of dioxin in the United States for an entire year are estimated by the EPA to aggregate to 30 pounds. But this becomes a possible source of effects on human beings only if it is eaten. So what is of possible concern is some small fraction of the 30 pounds annually nationwide (probably some very small fraction of one pound per year throughout the country) that ends up where it could be taken up by plants or animals used for food, and then in fact come to be eaten by people. The costs of further control would be high (since there are many small sources that would need to be tightly controlled), and the plausible effect on health would certainly be subtle and perhaps nonexistent. But since perceptions of dioxin are very strongly anchored in both danger and guilt, any threat from dioxin readily prompts a stark "better safe than sorry" response.

Since the costs of yet tighter control would be high and the risk subtle, it might not be worth spending so much money to cut so marginal a risk. But that is not the point of a "do no harm" assessment, so to put the argument that way would miss the point. Rather the cognitive merit of "do no harm" is precisely that it is not about trading someone's dollars (owners or customers of some enterprise, or taxpayers in general) for someone else's life. The point of "do no harm" is not directly to raise questions of whether too few lives are threatened too slightly to justify the expense of tighter controls. It is merely to ask whether we can, in fact, have reasonable confidence that tighter controls would more often save rather than cost the very lives they were intended to benefit.

This might seem to fit nicely with common intuition, which ordinarily is marked by strong reluctance to accept intervention that harms some people to help others even if *more* lives would be saved than lost (see Baron and his commenters [1995] for an extended recent discussion). But here, absent fungibility, and with framing and fairness effects that have fallen in line with that, the prevailing social sense is that anything that reduces dioxin must be right. ''Do no harm'' aims to open such commitments to reasoned discussion by introducing at least a minimal sense of fungibility. And it seeks to do so in a way that cannot reasonably be read (or easily misread) as about how much pain we should inflict on victims for the convenience of the rest of us. The ''do no harm'' proposal only supposes that if we are going to take the trouble to do something, that something in fact ought to be more likely to have positive rather than negative consequences.

8.7

But a realistic assessment for something other than a popular substance like saccharin requires that ''do no harm'' be built into the process in a way that insulates an administrator (or legislator) from attack for an assessment that fails to come to the immediately popular conclusion. That insulation could only be provided if the assessment is part of a general way of proceeding that can win support as fair and sensible. It needs to be backed by legally significant language. General ''do no harm'' language need not be anything more than a simple statement of what most Americans assume is the way regulations are already supposed to be made. But much more detailed rules would be needed to shield the process from ''better safe than sorry'' propensities that would otherwise push things toward just the one-sided assessments that ''do no harm'' is intended to avert. An instructive illustration is provided by the National Academy of Sciences (NAS) saccharin study (1978) already discussed. The study was commissioned by Congress in response to the huge volume of public complaints about FDA's proposed ban. Unlike the FDA, the NAS study was not constrained by the Delaney Clause. And, as directed by Congress, the NAS study makes an attempt to consider health advantages as well as health risks of saccharin.

But the NAS study did not escape the risk matrix ''better safe than sorry'' responses that, naturally, affected many participants in the study, the staff of the study (which in the nature of NAS studies, does most of the work), and also some sources of funding for the NAS. Anyone who reads the saccharin study will find ample illustration of a tendency to treat evidence suggesting harm from saccharin in a much more generous way than the evidence of benefits from saccharin. The assessment was heavily focused on the risk that saccharin could cause cancers. The possibility that a ban might cause at least some people to gain weight, and some of those to suffer strokes or other weight-related health effects, drops almost out of sight.

But that tendency to treat adverse evidence as sound unless proven unsound, and just the reverse for benign evidence, would be much harder to avoid in assessments done under more politically sensitive auspices and with respect to substances that are routinely seen and discussed in ways that take for granted that they are dangerous and guilty.

Realistically, heavily uneven assessments could be avoided only if a clear set of rules with some legal force made it hard to do that. Here the same considerations that prompted Breyer's proposal for an interagency corps of risk managers (7.1) argue that the rules be organized under auspices outside of any operating regulatory agency, with its particular commitments, political concerns, and habits of how things are done. In general, the merit of "do no harm" is that it seems a modest and reasonable reform as a general way to proceed. But that runs into strong cognitive impediments in particular cases. Hence the idea makes political sense only as a proposal about how rules of the game are going to work, as established from outside the often inflamed arena of particular cases. We would want the rules to be addressed in a context that favors a commitment to public health in the widest way. That suggests that design of the ground rules for "do no harm" assessments be convened under the auspices of public health professionals, rather than professional environmentalists or economists or politicians.[7]

7. The best place to manage a study with responsibility for proposing ground rules for "do no harm" might be the Office of the Surgeon General, which, through Republican and Democratic administrations, has on the whole maintained a reputation as committed to saying what needs to be said in the interests of public health, even if it is unpopular. In particular, it has maintained that reputation with those segments of the public most likely to be skeptical of anything like "do no harm." Unlike more specialized agencies, the surgeon gen-

8.8

Some salient issues for a ''do no harm'' assessment would be: how widely the notion of unintended effects should be spread, how to compare dissimilar risks linked to a common choice, how elaborate such efforts need to be, and who should bear the burden of proof.

The logic that has led us this far implies that the range of unintended effects that can usefully be covered by ''do no harm'' ground rules is contingent on how far including side effects of a particular sort seems reasonable and fair to a reasonable and fair-minded person with no special expertise. So as already discussed, what should be included in ''do no harm'' here is not the same as the equivalent judgment for a version of ''do no harm'' (risk-risk analysis) that is really a fallback from a full cost-benefit analysis.

Health effects whose relevance is disputed even among experts in cost-benefit analysis are not promising candidates for the ''do no harm'' assessment discussed here. We need a distinction between what can be seen, in general, as relevant effects, against what—even at the stage of agreeing on general principles for assessment—will be seen as debatable as effects (they might or might not be real), or debatable as relevant (they might or might not be unfair to victims). I will call those likely-to-be-disputed effects ''trickle-down'' effects, choosing an invidious term (as is appropriate for effects we choose to leave out of a ''do no harm''

---

eral has a broad mandate that covers every aspect of health; the office is not intrinsically focused either on particular health issues as specified in legislation or by the administrative practices that guide an agency like the FDA or the EPA. Nor does the surgeon general sit atop an establishment with entrenched political and organizational commitments that on the whole go against the very idea of ''do no harm.'' On the other hand, since the surgeon general's institutional commitment is to health, it is far less likely than an alternative like OMB to be suspected of undue zeal for economic efficiency. Yet some degree of attention to fungibility is intrinsic to the surgeon general's responsibilities, and more generally to the public health medical community. Resources for public health are far from infinite, so there must be some sensitivity to where spending would have important effects and where it would not.

assessment) but also one that conveys a lot of the sense that makes some effects seem merely trickle-down: ambiguity in where the effect is coming from and where it might end up.[8]

The most important example is imputed health costs of regulation. Paying for regulations takes money out of some pockets, usually, though invisibly, taxpayers or consumers at large. So people are left poorer, hence with less to spend on anything they might want to spend on, which includes health and safety. In general, diverting resources from more efficient to less efficient uses implies some decrease in spending on health and safety. And, indeed, "poorer is riskier" is very well documented.

But these imputed health risks are unpromising candidates for inclusion in the "do no harm" proposal discussed here, since these are indeed trickle-down costs, whose magnitude, and whose victims are likely to be disputed even among economists. After all, what are expenses for some people is income for others. And if an increase in risk brings income or satisfaction with it, it would be misleading to treat that compensated risk as if it were simply a burden. So not all side-effect risks belong in a "do no harm" assessment: there are costs which are not compelling on equity grounds even if a good case can be made on efficiency grounds (trickle down costs); and there are costs which are compensated, so that they are doubtfully relevant on either efficiency or equity grounds. And a "greedy" format that seeks to squeeze in effects

8. "Do no harm" relates also to Hans Inhaber's (1982) work on total risk assessment for energy systems. "Do no harm" would bear a family resemblance to that work, but also vary from it in important ways. The variance would turn on how far some health or safety consequence of a choice (perhaps an indirect effect, due to offsetting behavior, or a substitution effect) would be *felt* by reasonable nonexperts to be a fair component of an overall judgment. So Inhaber considers lives lost in the construction trades as part of the health and safety effects of various energy choices. But people in those trades see new construction projects as favorable, not adverse, opportunities. Hence counting risks from construction as if they were just like risks from the release of a carcinogen raises logical difficulty and, more important in terms of "do no harm," profound cognitive difficulties. On the other hand, risks imposed on third parties from heavy trucking operations connected with construction plainly should be part of a "do no harm" assessment.

that are in dispute even among economists would undermine the cognitive point of "do no harm."

8.9

On the other hand, various other health costs routinely ignored in regulatory decisions are obvious candidates for inclusion in "do no harm." For example, at low doses, it often turns out that a pollutant shows evidence of both positive and negative effects. So at low doses, a substance reasonably judged a carcinogen (it is associated with an *increase* in some form of cancer) will often turn out to be also reasonably judged an anticarcinogen (it is associated with a *decrease* in some other form of cancer). Then a direct consequence of reducing exposure in order to lower the risk of one cancer is to increase the risk of the other. When that is the situation it is hard to imagine any principled basis for ignoring the increase in risk that would come with tighter control and considering only the decrease in risk of some other cancers. And presumably reasonable and fair-minded people would think that a sensible process for choosing about risks would not ignore that. Rather, a fair procedure would care that people might be hurt as well as helped—and all the more so if, as is commonly in fact the case, ex ante the two groups are the same group.

But dioxin provides a ready example of how, in a concrete case, that might be easily overridden *unless* (which is the point of the "do no harm" proposal) a mandate for attending to both positive and negative effects had been firmly built into the ground rules for assessing such choices.

On the record (as of 1995) it appears that trace amounts of dioxin in fact do inhibit some cancers while promoting others. But, as has been discussed here often, a person seized with the intuition that it would be "better safe than sorry" to tighten regulations will easily think of some reason to justify not considering that possibility. We can expect arguments like, "How do we know that the studies were not corrupted by the chemical industry?" and "People feel concern about cancers caused by a positive act of pollution, not about cancers caused by *not* polluting."

There is no way to disprove the first point, since in principle it is impossible to prove a negative (that the study was not rigged or in some other way misleading). On the other hand, focusing exclusively on the possibility that the study is wrong—ignoring the possibility that it is sound—is hardly reasonable, however "right" that feels to a person in the "better safe than sorry" cell of the risk matrix. Nor is it credible to suppose that people really do not care about cancer that might be caused by regulatory intervention, only about cancer that might be prevented by regulatory intervention.

Note the cognitive effect of shifting the implied framing between the original argument about what people care about (two paragraphs back) and its rebuttal in the previous sentence: this illustrates yet again the cognitive effect, but also the contingent nature, of what might be seen as the status quo (5.10). But in the dioxin context of a substance "everyone knows" is dangerous and guilty, a claim that exposures actually encountered might have overall net benign effects—or even any benign effects at all—is a claim that of course you cannot believe, since you already have a very firm intuition that it cannot be right. Studies have followed the medical histories of people who were exposed to dioxin by a large release in 1978 from a chemical plant in Seveso, Italy. And indeed, the differences seen in cancer incidence include decreases in some cancers as well as increases in others, with a slight decrease overall (*New York Times*, 26 October 1993, C4). A parallel result from in vitro experiments on cells was discussed in *Science* (9 September 1994). Consistent with the argument here, the eight-column *New York Times* headline drew attention to increases in cancer; the overall decrease was relegated to a one-column subhead. The lead paragraph of the story followed the same pattern, as did the lengthy text. The on-line database through which I located the article provided an abstract that mentioned only cancer increase and ignored the overall decrease. Specifics on cancers that showed increase came high in the story. Specifics on cancers that decreased came only late in the story, though on the usual practice of news editors, at least one of the decreases might have been judged a good deal more newsworthy: women exposed to dioxin experienced only half as many breast cancers as those not exposed.

8.10

Suppose two risks are both invisibly small, but one is firmly anchored in public perception as dangerous and the other is not. Or, a variant of this difficulty, suppose one risk is exotic (hence is easily framed as something new, a loss from the status quo if accepted) but the other is familiar (hence most easily seen as part of what we accept). Unless as a matter of principle, many people would approve government intervention that would kill more people than it would save, provided the victims were killed in some way that would be hard to notice (as some merely invisibly small disturbance of usual events), then some effort makes sense to constrain the chances for cognitive propensities we cannot escape to distort perceptions in ways that eventually are likely to seem unreasonable.

There will also commonly be cognitive difficulty in seeing invisibly small and unintended side effects on the same scale with the focused risk, even when the focused risk is also invisibly small. A risk that has become the focus of regulatory attention, however microscopic, will almost inevitably be some risk that has come to be something that "everyone knows" is worrisome, thus it does not have to be visible to cause concern.

A procedural rule that would help on this matter and on several other difficulties would be that extrapolations from evidence to estimated effects should be by the shortest sensible route. Suppose we want to estimate the annual income of a group of 50-year-olds. Their income is not directly available but we have to estimate it from some information about their backgrounds. We could get statistically significant, and for some purposes even pragmatically significant, estimates from knowledge only of the years of schooling of the individuals. But we would get better estimates if we could extrapolate from their earnings at age 30. And we would get very good estimates if we needed to extrapolate only from last year's earnings.

Similarly we can expect more reliable estimates of health effects from observations at actual high exposures to human beings (as at Seveso) extrapolated down to regulatory standards than from extrapolations from enormous doses to mice (7.7). And even extrapolating from mice, an issue that needs to be taken very seriously is whether we could expect more reliable estimates of effects at the trace levels likely to be permitted under regulations from the lowest doses that produce detectable effects than from the highest doses at which the animals can survive long enough to develop cancers. In terms of conventional measures of *statistical* significance enormous doses produce (of course) far less chance that we are seeing merely random effects. But an extrapolation from the statistically significant effect of a dose equivalent to 800 cans of diet soda a day is not very plausibly a more reliable or more scientific way to estimate effects of saccharin on humans than an extrapolation from effects of necessarily more modest statistical significance of the equivalent of 8 cans a day. Rats dosed with the equivalent of 800 diet sodas a day were used to impute a bladder cancer threat to humans. But in the very same studies rats dosed with the equivalent of 8 diet sodas a day showed a substantial decline in overall cancer rates not only compared to 800-can rats but even compared to rats given no saccharin at all (OTA 1977, p. 57). A person who thinks the extrapolation from 800-can doses is more reliable can only be someone who misunderstands the significance of the term "statistical signifi-cance."

This issue, very plainly, is an important one for design of an assessment which takes seriously the question of whether a regulation promises more help than harm. It *commonly* turns out that (as in the saccharin tests) such effects as are seen at lower doses involve a mix of benign and adverse effects. So extrapolation from extreme doses involves a double compounding of pessimistic assumptions. As already discussed (7.8), that ignores strong reasons for concern that effects at enormous doses may be substantially artifacts of the vulnerability of animals under such abnormal conditions. Then ignoring benign effects at more realistic exposure levels compounds the bias towards pessimistic estimates. Logically, that multiplication of steps maximizing perverse effects cannot make sense if the intent of the regulation is in fact to "do no harm." Nor is it true, even if it were sensible, that only by exploiting extreme results can formal statistical analysis be brought to bear. Statistical techniques (meta-analysis, especially if combined with Bayesian methods) have been developed in recent years to formalize aggregation and use of individually marginal results.

How could it be that a highly toxic waste product like dioxin could decrease as well as increase cancers? The salient answer is that chemicals and radiation have been present (of course) throughout the evolution of life on the planet. Even novel synthetic chemicals always have some common properties with naturally occurring substances. Radiation from modern technology is identical to naturally occurring radiation. Dioxin is produced in forest fires. Man-made pesticides are a tiny fraction of natural pesticides evolved by plants. We can expect that over evolutionary time, living things will exhibit some success in exploiting positive potentials of what is encountered in the world as well as some success in controlling negative potentials (though sufficiently high exposures eventually must overwhelm defenses). Consequently, it is not a surprise that many things clearly toxic at high doses show a mix of benign and malign effects at low doses.

Hence a realistic "do no harm" assessment would exhibit concern that the conventional statistical significance of results at extreme doses was qualified by the large additional uncertainties introduced by attempting long extrapolations. And it would certainly not ignore what can be seen (which will often include benign effects) at more moderate doses.

8.11

If "do no harm" is to be effective, it needs to deal reasonably in another sense with marked variations in the nature of uncertainties at issue for different risks. Mundane hazards are characteristically data-rich hazards. At the other extreme, we have data-thin risks of (say) suspected weak carcinogens, which typically involve long-delayed effects, and extrapolation from limited data far from the regulatory regime. A "do no harm" assessment would often involve questions of how to provide a common assessment for data-rich and data-thin risks. We could not reach a reasonable judgment by summing over pessimistic estimates, since for well-documented risks (low ratio of variance to mean), pessimistic estimates will not vary much from most likely estimates. But for poorly documented risks (high ratio of variance to mean), a pessimistic estimate can easily vary by orders of magnitude from a most likely estimate. So across heterogenous risks we would easily get wholly unreasonable overall estimates by summing pessimistic estimates.

Special attention would be needed to assure that mundane risks (risks that are bundled into the sense of everyday life) are not simply neglected. And attention is needed to the consequences (mundane or not) of the physical activity involved in carrying out a contemplated standard, and also perhaps consequences of the reallocation of resources required to pay for that. So in an oblique way, dollar costs of compliance come into play here, since although the "do no harm" assessment is not about dollar costs, and not even (as discussed earlier) about potential "trickle-down" income effects on health of dollar costs, nevertheless dollar costs will usually point to places where there may be physical risk consequences that should be in a "do no harm" assessment, or reallocation of resources that effect risk.

Compliance with contemplated stricter standards for dioxin, for example, would cost U.S. hospitals $1 billion for upgrading their incinerators (to heat materials to a level sufficient to break up dioxin molecules.) Clearly, that must have consequences. The "do no harm" assessment would not deal with the dollar costs of compliance in a cost-benefit sense, or even in a risk-risk sense, but it should deal with the consequences of doing things differently. The cleanup of Times Beach provides a more transparent example. That cleanup involved trucking 500,000 loads of dirt. Clearly, there would be a calculable risk to third parties of "normal" trucking accidents from half a million shipments. Guidelines for a "do no harm" assessment should mandate attention to such mundane risks, not just exotic risks that more easily make headlines.

Is it possible that $1 billion in dead-weight loss costs could be imposed on hospitals (for incinerating dioxins) with no adverse consequences whatever on health? It is hard—in fact impossible—to believe that. So guidelines for a "do no harm" assessment should be such as to assure some reasonable effort to consider such effects.

Logically, the situation here is no different than for taking into account the possible weight effects of banning saccharin. But we have no sense at all (until someone has taken the trouble to investigate and report on it) about consequences of spending $1 billion on tightening incinerator standards in hospitals. Hence the default and anchor-and-adjust processes, which have been discussed at length in earlier chapters, operate to make the weight risk from a ban on saccharin easy to see as worth some attention, but make the perverse side effects of tighter controls on dioxin *seem* obviously negligible. But logically it is not obvious at all that those effects would be negligible relative to any plausible health benefits from tighter controls.

So (reviewing) both with respect to direct effects (on cancer incidence) of tighter dioxin standards, and also with respect to indirect effects (the health consequences of imposing a $1 billion burden on hospitals) even in as conspicuously difficult a case as dioxin, the first cut just sketched suggests that a serious "do no harm" exercise would be likely to yield an assessment very different indeed from what will come from a black-and-white "better safe than sorry" assessment.

8.12

Who should bear the burden of making a case, those who urge precautions or those who are skeptical that the precautions are needed? Usual practice is to put the burden on those who wish to change the status quo. And one could argue that it is the precautions mandated that create a change from the status quo. But cognitively it is the activity that creates or might be creating a risk that will usually be seen as disturbing the status quo, and proposed regulations as only a response to that. Since the point of "do no harm" is to design a procedure that will seem fair even when "better safe than sorry" intuitions are strong, the burden of initially going forward with a "do no harm" assessment probably should fall on those who want to resist a regulation or precaution.

But that initial case would require a response. That response from proponents of caution would not require anything like *proof* that net benefits would be positive. But a *plausible* counterargument would be needed, meeting language like that set out in the opening paragraph of this chapter, and commensurate with the strength of the "do no harm" objection to the regulation. It should not be possible to evade the obligation (if challenged) to make a plausible showing.

8.13

Following Shweder and Haidt (1993), we can note that in all societies moral intuitions appear in three realms:

1. What is fair or just for individuals? This is a notion that gets exceptional weight in modern democracies but is never entirely absent in any culture.
2. What is just with respect to things deemed sacred in a culture? In our own culture this clearly includes a concern about what is just with respect to the natural world, to life on earth, and to future generations.
3. What would be fair and just principles for how society governs itself? What will make for a wisely-governed society?

Of these three sources of moral intuition, items 1 and 2 are the natural focus of "better safe" concerns. What causes moral concern plainly includes concern that innocent individuals may be harmed, or that sacrilege will be permitted. But it also provokes a sense of moral wrong to knowingly make choices in a way that violates what makes sense as general principles of how to choose (item 3). That kind of concern would appear as some variant on the Kantian theme: if we always did what we are doing in this case, would that be socially sensible? If not, should we not choose better rules about how to choose?

So the third of the moral categories invites attention to choosing the rules for choosing about risk. "Do no harm" is a deliberately modest proposal (compared to what full cost-benefit analysis would require.) But it has the very significant cognitive virtue that it does not immediately violate the "better safe than sorry" sense of things. It only broadens the focus of what a fair process would attend to by way of seeking to be better safe than sorry.

Logic is not irrelevant to cognition. But as I have been stressing, it is realistic only in very special cases to suppose that logic governs cognition. An effective ''do no harm'' process cannot be such that it would be seen as an unrestrained slap at ''better safe than sorry'' intuitions. Rather, ''do no harm'' only would build into the process a check on how far a choice really is more likely to make us safer than to make us sorrier. Logically, that would be a very modest step indeed. Pragmatically, it might be very substantial.

Political Externalities

9

In 1970, at the very outset of contemporary environmental legislation, a set of numbers for auto emissions took on a sacred character that made any questions seem on the edge of obscene. A look at the legislative and political process that wrote those numbers into law on a very tight schedule finds no one in the process who seemed to care about the costs that would be imposed by very tight standards mandated on a very tight schedule. (The key actors were President Richard Nixon and the then-leading prospect for the Democratic nomination and chair of the committee handling the bill, Senator Edward Muskie.)

No one, also, seemed to care that the numbers acquiring this almost sacred character had in fact been pretty much picked out of a hat. The bureaucracy had been ordered by the White House to produce, over a weekend, some numbers the administration could set out as long-range goals for the auto industry. So we had numbers picked out of that hat with no awareness that political conditions (in particular, the competition between Nixon and Muskie to be more forthcoming on this suddenly salient matter) would be such that these numbers would be written rigidly into law on a much speeded-up schedule. The legislation passed the Senate with neither an hour of hearings nor a single dissenting vote: but a great many hours of debate would take place over the next twenty-five years, and (as this is written) the standards have yet to be fully met.[1]

9.2

We had passage of draconian legislation without even a pretence of serious discussion. A logic to how that occurred emerges by noticing a close analogy with a fundamental issue for economic theory. The underlying structural difficulty is a political analogue of the economic problem of externalities. Governments must act to control auto emissions because the harmful effects attributable to any single car user are inconsequential. Significant air pollution is due to the combined effects of very large numbers of automobiles. So even a public-spirited individual driver has little motivation to incur any significant cost to cut his own emissions. Ordinarily he cannot see, nor can anyone else see, any effect of a cleaner engine on his own car's exhaust, much less detect any drop in the total level of pollution in his community. So the pollution from his car is an "external" cost—as economists use the term, a real cost that the individual is not motivated to take into account when deciding how to use resources.

Since the consumer is inadequately motivated to take into account the costs of pollution, the firm making what the consumer buys is even more weakly motivated. For it profits by providing the consumer with what he is willing to pay for. The result is a social dilemma in which individual consumers and individual firms have no adequate incentive to take pollution costs into account, although all have a joint interest in having everyone do so. The solution to this dilemma is to have an agreement that each will do her share on the condition that everyone else also does so. Governments provide a mechanism for arranging and enforcing such social agreements.

1. For details see my 1977 article, "The Politics of Auto Emissions." This section is taken mostly from the conclusion of that article. In a formal sense, hearings had been held, since the bill was technically a revision of a far milder bill originally on the table. But in fact it was a new bill whose key provisions had never been considered at the hearings.

A political analogue of this dilemma arises when the interests brought to bear on a political choice do not in fact adequately represent the interests at stake. That is discussed a great deal, and sometimes reasonably effective steps are taken to lessen or balance the advantages of what we familiarly refer to as "special interests." But a peculiar form of political externalities arises where the lack of balance is hidden because an appropriately full range of interests *appears* to be represented, but in fact is not. And that problem arises very conspicuously in environmental controversies. The number of firms making a product is always very small compared to the number of buyers of the product, so it is easier to regulate the producers rather than the consumers. It is also politically easier to regulate the producers, since what the regulation costs the consumer is then hard to see. So the direct interests in the legislation (or the regulations coming out of an administrative proceeding, or the lawsuits that shape policy) become the environmental interests versus the producer interests. The consumer interest is not directly at the table. Finally, since the consequences for an individual are slight and the chance of an individual influencing the result is slight, few people would be well motivated to work at following the arguments in detail other than those with strong commitments (say, pro-environment or anti-government regulation), whose minds are already made up.

Out of such circumstances an important externality will easily arise that is political, rather than economic, in origin, *and* which is hidden in the sense that an important imbalance is in operation but cannot easily be seen. In the case of auto emissions, the environmentalists perceived their role as maximally protecting the environment; if industry had perceived its interest as minimizing total cost, then out of this clash of competing interests some reasonable social balance might have arisen.

But in fact industry's interest was not to minimize total social costs, but to minimize those costs it was forced to bear—which did *not* include much of the costs of emissions controls. For as long as all makers had to meet the standards, market conditions would essentially assure that the costs were passed on to consumers.[2] We had a context in which social perceptions were conspicuously anchored in danger and guilt: there was a strong "everyone knows" sense that auto emissions were bad, and the automakers indeed had a bad record when it came to recognizing that emissions controls would make sense. In that context, a direct attack on the now almost sacred numbers could be judged most likely to merely anger people or raise even further the mistrust of anything industry representatives said. Pretty clearly, that was the judgment of industry lobbyists, since as described in some detail in my 1977 article, the industry did not directly challenge the proposed standards.

But whether the standards indeed were excessive is not essential for the argument here, though there is a good case to believe that they were.[3] Whatever the best judgment might be, there is surely something odd in the passage of legislation with costs of billions of dollars annually without any serious discussion of the merits of the standards chosen as against some alternative.

The salient counter is to note that, in the environmental cases discussed here as in a variety of other recent matters (crime and drug legislation being conspicuous), the public is getting what it wants, even if experts are distressed at what they see as the extravagant irrelevance or perversity of some of the things the public wants (such as the "three strikes and you're out" provision in crime bills). For a segment of the electorate too large for a political figure to risk offending, "three strikes" has become a litmus test for whether a person is serious about crime. On the other hand, there is also a good deal of evident mistrust of government from a sense that the government wastes taxpayers' money on programs that do not work. So it makes some sense to consider that things might work better if the process could be reformed in some way that makes it less likely that there will be politically irresistible public support for things that in fact will not work.

2. Exactly how that would work out would depend on the market situation (price elasticities), and conceivably, tight controls could actually be profitable for the makers, as more intuitively might be the case if a law was passed requiring all cars to have air conditioning.

3. Standards for U.S. automobile makers are about five times as strict as those in Canada or Western Europe. See also Krupnick and Portney (1991).

What I am calling "political externalities" is not an unrecognized problem. But the kind of case usually recognized is where some interests are not at the table at all or are ineffective because those claiming to represent broader interests lack the financial or organizational resources to be effective. But the particular focus here is on cases, which are peculiarly common in environmental matters, where on the surface it seems that an appropriately full range of interests are well represented, but where that turns out to be illusory. So we are concerned here with cases in which political externalities can take a form that is particularly difficult to deal with, since their existence is especially hard to notice.

But a sophisticated analysis of the broader issue, to draw on another useful notion from economic theory, requires concern about "second best" difficulties. Reducing the influence of special interests does not guarantee better social outcomes. And in particular, it may not do so at all if broadening the discussion effectively makes the process more vulnerable to a political externalities problem of the special sort that is our focus here. As usual in politics, good results are not assured by good intentions. So the problem of dealing with the broader issue has a subtle but often important link with the special problem of hidden political externalities that is the focus here.

9.3

To illustrate, consider what happens in the context of public participation in local environmental decision making. There are three basic situations, which I will call "balanced," "unbalanced," and "suppressed." The characteristic *balanced* situations involve land use conflicts in Western states that pit economic interests against wilderness conservationists. Representatives of both sides are brought together with public officials in meetings managed by a new sort of professional mediator ("facilitators") specializing in getting people representing diverse interests to sit down and over a sometimes long series of meetings, gradually talk out their conflict. A basic though never explicit tactic appears to be to let the zealots on each side talk enough to give the moderates on both sides a sense that they have more in common as human beings with their moderate adversaries than with the zealots on their own side. These tactics sometimes work and sometimes produce agreements that moderates on both sides see as clearly better than endless fighting in the courts and elsewhere. So such efforts have often turned out to be well worth what they cost in time and money.

But in these land-use cases we have good prospects for actually having at the table effective representation of the interests at stake. The participants collectively represent what people across their state care about. So if they eventually can work out an agreement, and professional facilitators can help manage that, it is likely to be something that looks reasonable as a pragmatic political compromise of conflicting interests. No essential consideration for how to handle these land-use disputes is likely to be ignored in discussions among ranchers, environmentalists, and public officials—in contrast to the absence of serious attention to overall social costs and benefits that marked the auto emissions case.

But consider an unbalanced case. Suppose the issue (as at Hanford or Rocky Flats, Colo.) concerns how to clean up facilities left over from the Cold War. And suppose, as is now routinely the case, that the voices at the table are environmentalists, people from the surrounding communities, and representatives of state, local, and tribal government. So the list of participants looks pretty much like a list of participants for the balanced cases of land-use disputes. But here it would be a serious mistake to suppose that this list covers a rich enough range of interests so that, if an agreed position can be reached, it too would be a reasonable political compromise. For these cleanup controversies, in contrast to the land-use controversies, that is not likely at all.

For suppose some piece of the cleanup effort may be completely worthless, as appears to be so for treating well water at Hanford as contaminated waste (8.3). State and local officials, and contractors carrying out the work (who will not neglect their financial interest), all have a strong economic interest in a large cleanup effort. Exactly as with a military base that the Defense

Department would like to close, the cleanup provides jobs and income. Some people are worried about risks of environmental damage. But others are worried about losing jobs or contracts.

These worries complement each other. There is rarely conflict between people worried about one and people worried about the other: rather, if you are worried about one, you find it easy to share the concern of people worried about the other. Local people have no interest that conflicts with "better safe than sorry" precautions. More cleaning up means both more jobs and less worry, no matter how faint the latter is. The only incentive to make any serious effort to question trucking well water across the countryside to be decontaminated before being thrown away would arise because sooner or later even the most extravagantly funded cleanup runs out of money.

Sooner or later, consequently, a budget constraint must enter that will focus attention on eliminating some projects. But how far that will prompt elimination of particularly ineffective projects is highly doubtful. There is no one in the process who has a serious incentive to ask whether the cleanup overall makes reasonable sense, as against the possibility that a cleanup effectively as good could be done with half the budget. There is no place at the table for concerns about how far money is being spent on projects with no discernable effect on the well-being of either people or the environment.

What about "psychic" benefits? People who were worried feel protected, and so feel better even in a case where it was really true that there was nothing to be worried about in the first place. The topic of psychic benefits deserves a much more detailed discussion than I can give here: concerning, for example, the extent to which psychic damage can be created by a process that might reward people if they are suffering such damage. The cure for the damage is then a contributor to the extent of the damage to be cured. And one would like to see some evidence that responding to purely psychic damage in fact yields long-term benefits even to those provided this treatment. In the short run, no doubt, there is an effect. But some disillusionment, with its psychic costs, surely might arise when those who were suffering find nothing is really different (since nothing could be really different if the harm was only psychic to start with). If social intervention (favoring inoculations, or seat belts) actually saves real injuries, not just psychic injury, we expect to see concrete evidence of that in fewer people dying. Is there evidence of any such bottom line benefits from response to psychic injury? And there are other significant questions to ask before a policy of catering to imaginary injuries is taken to be sensible public policy. And that is so even though the pain can be and indeed usually is perfectly real even when the injuries are only imaginary.

But returning to the cost constraints alone, suppose the budget is large enough to accommodate a large fraction of the effort on projects that might well be worthless (no one being motivated to probe that possibility very hard). Where would an incentive come from within the interests at the table to urge not wasting money on such projects? "Better safe than sorry" will rationalize anything. The money effectively comes from heaven (i.e., from taxpayers at large) and choices have no noticeable impact on the taxes of those at the table or specifically represented at the table.

We have, then, the problem of political externalities introduced in 9.2, and really a peculiarly perverse form of that, since those at the table understandably see themselves as performing a public service. Most, indeed, are concerned citizens volunteering their time. No one is doing anything wrong, and most are trying their best to do what is right. The problem is not in the behavior of those at the table, for even those who are not volunteers are usually serving a legitimate interest they openly represent. But a large interest is now missing: someone whose responsibility is to society at large and its interest in reasonably effective use of what (after entitlements, interest on the debt, and national security are taken care of) is a disturbingly modest supply of public resources for everything else that needs attention and funds.

9.4

Yet we might suppose that since federal officials must also be part of the process, that should balance out such unbalanced cases: the party concerned that things make sense from a national view would be provided by the federal officials. But the case we started from (emissions controls on automobiles where everything was done from Washington) shows how far from correct that reasonable-sounding surmise might be. In that auto emissions case, the process was predominantly in the control of individuals who could have been expected to have a national perspective, with concern about overall costs and benefits as a central part of what they took to be their responsibility.

And no doubt making sense from a broad social view was something participants cared about: but not enough, and with enough leverage, to make that concern effective, given other things the actors involved were moved to be concerned about. Politicians depending on votes and environmental organizations depending on contributions found they cared more about how their performance would look to their constituencies. Managers of the auto companies cared more about how what happened would affect their ability to handle their jobs and enterprises.

Breyer's "last 10%" problem (the economists' diminishing marginal returns problem) was no one's problem—or least not so saliently within the responsibility and authority of any individual actor to make that argument one that anyone wished to take on. For as usual on such matters, and as discussed repeatedly here from chapter 4 on, "twoness" was very much in evidence. What had been hastily contrived, substantially arbitrary numbers quickly became a bright line separating safety from danger—indeed, separating good from evil. Questioning those numbers was inviting any adversary to question your integrity. In the circumstances, everyone saw other aspects of the problem as what they really cared to attend to. No one was motivated to challenge what "everyone knows" was right, which was to be "better safe than sorry." Any risk, even any speculative possibility of a risk, should not be inflicted on innocent victims.

This kind of situation yields what I mean by the "suppressed" case. One might have supposed that if the standards were unreasonably tight, the auto industry would take on that fight, given that tightening of controls on auto emissions would cost billions. But those costs (as everyone involved—government officials, legislators, environmentalists and of course the industry itself—fully understood but no one cared to say out loud) would not be paid by the industry, but by the pubic at large. Necessarily so, since the cost of controlling auto emissions far exceeded the total profits of the automakers, so who else could pay them? And properly so, since who creates auto exhaust fumes but drivers of autos? So the industry focused its concern on fine print in the legislation that did not involve such unpleasant public relations difficulties and, consequently, such poor prospects of receiving a sympathetic hearing. And everyone else could comfortably suppose that if the standards deserved questioning, it was the industry's role to do that.

9.5

For auto emissions, little public spending was involved, only regulations that would require billions of private spending. But environmental cleanups (most obviously where the source of what needs to be cleaned up is a government program) commonly require public money. We might have supposed that federal officials involved with choices about cleanup efforts could then hardly escape concerns about that "last 10%" problem. It would require federal dollars for essentially all the work. A broad social interest in whether spending is likely to be effective or wasteful is even more conspicuously at the table than in the auto emissions case.

But that does not assure that broad public interest will be *spoken for*. The easier course for public officials is usually to agree with whatever is proposed locally and promise to work for getting as much money as possible from Washington to pay for it all. Eventually the buck must stop somewhere, but that control gets exercised at the politically safer level of government-wide budget constraints (at the Office of Management and Budget [OMB] in the executive branch or in Congress by the budget committee), after layers of promises from officials who can then disclaim any control over what happens at these higher levels. So at the local level of dealing with cleanup at particular sites, an agency does not need to do what is politically unpopular: it can promise to spend as much money as Washington provides, and the express distress when powers beyond its control do not provide everything that was agreed to.

But those constraints have not been so severe that along with large amounts of environmental spending that gets good results, large amounts of money (public money, and through mandated regulations, even larger amounts of private money, ultimately paid by consumers at large) are also spent on projects whose good effects are unclear—indeed whose good effects may not exist, or even, as stressed in the discussion of "do no harm" in chapter 8, whose net effects on health and safety can be most plausibly expected to be negative.

That this in fact is widely understood is indicated by the overwhelmingly positive response and marked ineffectiveness of attempts to generate opposition to the Breyer nomination to the Supreme Court. Breyer's book was published only a few months before his appointment, and of course it is outspoken on this matter of extravagantly misdirected environmental regulations and spending. But the absence of difficulty due to that, in contrast to the killing opposition to another nominee (Robert Bork) whose published views went against popular sentiment, surely reveals that many people who would not want to say so out loud thought Breyer right to say what neither the president who nominated him nor the explicitly conservative, explicitly probusiness presidents who preceded him ever found it politically prudent to say. But consider the situation of an EPA or DOE official sitting in on a public participation meeting, and probably also when appearing at a Congressional hearing, who spoke bluntly about the sorts of misallocations that bothered Breyer. Sounding like Breyer would be highly risky, since it would be likely to so offend the public mood on these matters as to destroy the official's effectiveness as someone that participants would listen to. She would probably have to be replaced by someone who would not do that.

Reluctance to give offense is peculiarly strong when, as in fact is commonly the case, the federal agency is seen as a guilty party.

A guilty party's view of what his punishment ought to be does not carry much weight, and an attempt to imply anything different easily gives offense. Indeed, a main point of the public participation exercise often appears to be to show how much the federal government admits its guilt. By freely confessing to past sins, it is hoped, but seems to me likely to be only futilely hoped, officials might win the future trust of the public. And in that context, federal participants are not likely to be encouraged by their superiors to say anything strong about concerns that nothing to speak of is being obtained from large outlays of tax dollars. Caution and humility is the way. In contexts where federal officials have no choice but to defend a federal program, since it is an on-going program, not merely the relic of a now-discontinued program (such as making plutonium), the federal official's lot is not a happy one.

A common result is that in response to worries about some matter that by any reasonable standard has already been attended to, a further study or further work is ordered. But real—not to mention imaginary—imperfections can always be found in any such effort. And the very fact that the study or work was undertaken becomes a demonstration that a real problem is there: or else why would that money be spent? So there is yet another reason to distrust the official view.

In general, then, it is a vain and even an unreasonable hope that in public participation discussions about environmental matters—cleanup of defense sites, or decommissioned nuclear facilities, or chemical waste sites, and others less conspicuous—participation of federal officials would assure that someone would speak for concerns about extravagant use of public money, much less that such concern would be shown about money that might be claimed from private parties.

9.6

Even with present assessment procedures (in particular, even without the "do no harm" assessment proposed here) more could be done to provide an effective voice at the table whose mandate is to be concerned that tax dollars are not squandered, that efforts to improve health and environment should be effective and not misdirected, that large social costs should not be incurred for ephemeral benefits.

What seems to be needed is a voice insulated from the political and public relations difficulties that inhibit government officials from fulfilling what might be seen as directly their responsibility. We would like someone in the process whose own special responsibility is to worry about whether taxpayers are getting something of reasonable value for what is spent. Or, what amounts to the same thing, someone whose role is to worry about how far society in general is getting what it thinks it is paying for. We all have an interest in that, since nearly everyone pays taxes, and those who do not also have that interest since those who do not pay taxes are principally those unfortunate enough to have to rely on what the rest are willing to allocate from our own taxes to help them. Relative to other advanced industrial countries, we are not very generous. And wasteful spending not only detracts from that modest remainder but reinforces doubts that government spending can accomplish anything very useful anyway.

But while it is easy to argue for a voice—a kind of ombudsman for the suppressed concerns—it is hard today, as it already was two decades ago, to see how some pragmatically effective and feasible new actor or new institution might be established.

As possible models, we might consider parallels to the special entities created to insulate such matters as military base closings and tariff appeals. A minimal step might be to invite individual volunteers to participate in discussions of cleanup budgets, with the explicit assignment of representing citizens in general. That role would include a right to speak at meetings and to append comments to the recommendations reached by a public participation exercise. Even so, it would be hard to make that voice effective. But the minimal step of at least providing for someone representing citizens in general (including but not limited to their concerns as taxpayers) as an explicit part of the process does not appear hard to accomplish, even for an agency with serious public trust problems. That would be particularly manageable if an oversight agency such as OMB or the Congressional Budget Office asked that it be done.

9.7

What I have been calling hidden political externalities occur when a logically present interest in fair and efficient use of social resources is effectively suppressed. The account here of how it comes about that broad social interests are sometimes effectively suppressed derives from the account of how fungibility is missed in cases where logically it is plainly present. Then a "better safe than sorry" response is single-mindedly in place even in a situation where, logically, there are other things to be considered, including, usually, other things to try to be safe about. Somehow, and I have tried to spell out how, participants are ignoring what in other circumstances they recognize as clearly significant, and treating as significant what in other circumstances they routinely treat as negligible.

The "do no harm" assessment proposed in chapter 8 is modest. It is only intended to open that closed door far enough to allow the "better safe than sorry" response to at least cover other things to be safe about. That is all it does, though as I have suggested, the consequences would be more far-reaching. And the prospects would be enhanced if "do no harm" were combined with other institutional reforms of the sort described in the previous section. The negotiator, or the ombudsman for us all, or the blue ribbon commission, or the interagency risk assessor would have a considerably easier task if a "do no harm" assessment were routinely built into expectations of what a socially fair analysis of risk would consider. Conversely, if institutional arrangements provided for a voice at the table committed to taking account of fungibility, then a "do no harm" assessment would help that voice play a significant rule: it could not so easily be seen as merely an intrusion on concern for victims.

So the discussion of political externalities in this chapter and of "do no harm" in chapter 8 are closely related. If not for the problem of political externalities, then an explicit "do no harm" assessment would not be so necessary. Without a commitment to "do no harm" assessment, other efforts to overcome the political externalities problem might have little chance to be effective.

Afterword

10

The same expert/lay conflicts that raise challenging public policy questions also provide an opportunity for seeing something of how cognition might be governed by habits of mind. I mentioned in the introduction that this study was motivated not only by concern with the public policy issue, but also by interest in more general questions of persuasion and belief. Concluding remarks on that broader theme are in order.

As in the studies on which this one was built, the argument here has been closely tied to evidence from toy puzzles that yield cognitive illusions. A common claim is that such anomalies are not really important since they occur under conditions that are far from what is typical of the situations people actually encounter in the world. And it is certainly true that we are not routinely blind to modus tollens inferences (as a naive extrapolation from Wason's "selection task" would imply), or routinely blind to the significance of base rates in cases where they are clearly relevant (as in Kahneman and Tversky's "cab" problem).[1] In general, we are not routinely vulnerable to gross cognitive illusions in the situations we actually encounter in ordinary life. So it is reason-

---

1. On modus tollens (the inference from "if A then B" to "if not-B, then not-A"), there is a huge literature, mostly focused on Wason's "selection task"; See, e.g., my detailed discussion in chap. 7 of *Patterns;* see also Evans (1983). On base rates, an extended review and discussion are found in Koehler (1996) concerning Kahneman and Tversky's "cab" and many related experiments.

able to deny that work on cognitive anomalies can be directly translated into generalizations about how people ordinarily cope with the world. But that claim itself is surely misleading if taken as warranting dismissal of cognitive anomalies as mere laboratory toys or mere oddities elicited by tricky questions.

It is commonplace for phenomena to be first clearly recognized in exceptional or artificial settings where effects happen to be starker than ordinarily encountered. And it is also commonplace for novel phenomenon so recognized to be dismissed as just a laboratory toy. That response is not at all a peculiar reaction to cognitive illusions. We live in a technological culture where scarcely a moment goes by when we do not, in some way, rely on features of how things work in this world which were originally recognized in some uncharacteristically simple situation. But a feature of how things work that had been previously unrecognized will easily look unimportant. Our experience in the world is (for such cases, by definition) that it has no noticeable consequences, or at least no consequences that have been noticed outside the odd context in which it is first seen.

So it is not surprising, and we can in any case observe from many such cases, that some knowledgeable segment of opinion is likely to see the novel claims at first as probably mistaken, and if that has become untenable, see the claims as of no practical significance. The point applies about equally to basic phenomena and to early "breadboard" technology. The steam engine, flying machines, and the telephone all provide technology examples. And of course a dismissive response is even more readily available for novelties not even worked up into a device (such as nuclear energy or electrical induction). The tendency, for all of us on at least some occasions, will be to dismiss unexpected phenomena as probably illusory and, even if not illusory, then of no practical significance.

So there is nothing novel about claims that cognitive illusions are merely laboratory toys or artifacts of trickily framed questions. That is a usual story, but a story that I think can be expected to encounter a usual fate. And it is also in no way surprising, or embarrassing for a view that sees cognitive anomalies as important levers for understanding human cognition, that *indiscriminate* generalization from illusion-generating puzzles would lead to serious understatement of human ability to approximate carefully reasoned judgment. Under conditions of normal experience, we do not see (or mainly see only in transient and inconsequential forms) the stark illusions that can be so strikingly elicited as responses to certain simple puzzles. If we *only* had to make judgments about familiar matters where intuition is well-tuned to experience, the dismissive claims would be warranted. But we don't.

We also make judgments, and judgments that at least collectively are very important, in contexts which are *not* those dealt with routinely. Then experience does not provide us with close feedback from similar cases that trains us to do what works and avoid what does not. And then there is no obvious reason to suppose that our ability to reach good judgments will be reliably better than we display in dealing with simple puzzles. To suppose otherwise really amounts to supposing that evolution provided us with special cognitive propensities for generating illusions in unimportant contexts. Surely it makes more sense to suppose that illusions reflect the operation of ordinarily effective processes inappropriately triggered in contexts outside the range of usual experience. The examples easiest to study involve simple (hence usually artificial) puzzles. But the natural occurrences are surely not limited to cases that are mere toy puzzles, with no important consequences.

On the range-of-contexts argument (1.5; so I won't repeat that argument here), *outside* the range of normal experience, such as easily occur in making social and political judgments, or even scientific or technical judgment beyond what is already reasonably familiar territory, we can expect that vulnerability to illusions will at least sometimes be a serious matter. In *Paradigms*, I tried to show how logically inappropriate but cognitively understandable responses in fact seem to have had striking effects even among the highly sophisticated, intensely concerned actors in various famous episodes in the history of science. And there is a strong case to be made—a compelling case, in my view—that anomalous cognitive responses are in fact common in social and political situations: meaning responses that seem to defy the logic of a situation, given the interests actually at stake. When we consider historical cases, such as the Salem witch-hunts and many other such episodes, that claim is not even in much dispute.

10.2

Yet even for the witches of Salem, and far more for contemporary cases, it is always possible to tell some sort of story of how the belief at issue in fact served (not defied) interests and logic. Any actual situation will be open to some uncertainty about the facts of the matter. And a complete logical analysis, even taking facts as given, will be very complicated compared to what is involved in a toy puzzle. The consequences of various choices would always be open to some dispute. Sometimes it is possible nevertheless to get a usefully wide consensus on what is reasonable. But we can never expect absolute agreement or airtight proofs. One virtue of studying toy puzzles, is that at least here things can be made simple and stark enough that we can sometimes see beyond all reasonable doubt when stories told to rationalize a judgment, however sincerely and earnestly told, are nevertheless illusory. And there are insights that then become accessible about the conditions under which striking illusions take hold.

It is instructive to note how readily an intrinsically simple but illusion-provoking puzzle like "chips" will prompt even highly intelligent people into defence of what is unambiguously the wrong answer (for chips, the 1/2 response). And it is also instructive to notice how stubborn that defense can be, once provoked. Just because "chips" is merely a toy problem, susceptible to a simple and clear physical test of what probability is actually right, even the most confident defenders of the 1/2 response sooner or later give up their commitment. But in many cases, coming to see what in fact makes defensible sense comes remarkably later, rather than sooner. And even once the simple logic of the situation has come into sight, there is still often a residual claim that in some way it was really the question, not the 1/2 response, that was wrong.[2] And even after those objections too have faded, for most

2. The residual response takes two forms, both utterly illogical so far as I can see, but from experience I can assure readers that both are sometimes pursued by highly intelligent subjects. One residual is to claim that the question is somehow intrinsically unfair. The question, for this sort of residual defense, should have been framed in a way that made the 1/2 answer the correct answer, and it was trickery not to do so. The other residual is to claim that the question

of us there is still a stubborn residual tendency for the illusory intuition to pop up, even though by now we know it is an illusion. As with the Muller-Lyer arrows, knowing that what we seem to be seeing is an illusion does not always overcome the propensity to see things that way. So in a case where things could never be so clear as they eventually become for the chips puzzle or for the Muller-Lyer illusion—which is to say, for almost any substantive focus of conflicting intuitions—we certainly have to allow not only for the possibility of illusory intuitions but also for the possibility that the illusion will prove very stubborn.

What made the expert/lay issues of this study a promising focus for close analysis was that it gave us a set of cases where more nearly than in other real-world cases, though not with the absolute clarity of a toy puzzle, we can see firmly held, thoughtfully defended intuitions that happen to be wrong. And then exactly the argument commonly made to deny the pragmatic importance of cognitive illusions (that people may be vulnerable to illusions in odd situations but not in the circumstances of normal life) implies that indeed we are quite possibly seeing illusory judgments in such cases. For in these expert/lay controversies the main division is over questions within the range of normal experience for one set of actors (experts, who by definition have a lot of experience in dealing with the issues involved) but well outside the range of normal experience for the rest of us. So we can see cases where experts, aside from a minority with substantial prior commitments, see things one way and lay judges without strong prior commitments see things in the opposite way.

The analysis here has yielded an account of why such judgments are often illusory. We can often, as illustrated in chapter 6, give a detailed account of how the illusion arises in particular cases, and of what steps helped dispel it, when that in fact occurred. And that can be done (almost as clearly as for something

---

as finally understood is correctly answered as 2/3, but as originally asked, the right answer was 1/2. And that can be maintained even when the original question was posed in writing in a way that by any normal standard was wholly unambiguous.

as simple as the chips puzzle) while restricting the invidious label "illusion" to cases where most of those starting from a clear intuition one way eventually came to see that same clear intuition as mistaken.

10.3

Of course, the argument I have made about expert/lay controversies has to be regarded as only an argument. It would profit from more detailed support than this one study could reasonably claim to provide. On the other hand, it seems to me that the study does open the door wider to an important range of issues, having in common the possibility that clear intuitions can be wrong—and wrong in the strong sense that the intuition comes to seem recognizably wrong to just the same sort of people (though for various pragmatic reasons not usually to the exact same people and especially not to every one of those people) that earlier were seized with a conviction that they could not be wrong. We can see in a range of concrete cases that clear intuitions about issues a person feels very strongly about need not be sound intuitions by the person's own standards of what constitutes a sound intuition. But that should not be a shocking claim. Surely everyone must realize that a *feeling* of certitude about some matter is not the same thing as certitude.

In these expert/lay cases, we have real-world cases that yield responses that approximate those of illusions of judgment. Once the illusion is finally overcome, an irreversible change in judgment occurs. Once a person comes to see that the right answer to the chips puzzle is 2/3 not 1/2, it never happens that the person changes back, just as a person who has once learned calculus or learned to ski a parallel turn does not revert to a state in which the logic of the calculus seems impenetrable or the parallel turn seems impossible. A person may still sometimes see the illusory intuition. But he no longer believes it might be right. Transitions almost that stark and irreversible seem to occur in various cases noticed in chapters 6 and 7—once awareness of fungibility has been prompted, even though that fungibility was always logically present, though somehow cognitively invisible.

The analysis of risk intuitions provides us with detailed cases in which intuitions clearly important in moral and economic value are governed by covert habits of mind, in a way that leads in these cases—although obviously not in every case—to perverse results. And these results are judged perverse by the very standards of people who originally saw those intuitions as incorrigible. That should go some way—in my view, it should go all the way—toward answering residual doubts that people can sometimes respond to situations with intense conviction but in ways that conflict with their own interests in that situation. For it is a good general rule that anything that exists is possible. And although Pascal was certainly right to say that the heart has reasons that reason knows nothing of, it is also the case (as I have argued) that those reasons will sometimes be very bad reasons.

10.4

Does that mean that we ought to abandon intuition as the basis of judgment? But most definitely that is not what it is implied by the argument here. It is not even what is possible on the argument here. Generalizing the "statistical" versus "visceral" argument of 2.11, at bottom *all* judgment is visceral. We *feel* an intuition or a belief is right. No one ever *decided* to believe something based on a logical analysis of what we ought to believe (as a computer analogue of forming a belief might work). We *find* we believe something, sometimes even in defiance of what we can conclude from mere logic. If challenged, we might later find that on reconsideration we do not believe the challenged claim after all. But this too would turn on *finding* that what we believe has now changed, not on a *decision* to change a belief in the light of evidence against it. On other occasions, we will find that we still believe what logically we do not know how to make coherent. But when that happens, as it sometimes happens for all of us, we do not *see* that as in defiance of logic, even when it most clearly is so. Rather, we feel a conviction that there must be something wrong with the logic that says we cannot reasonably believe that, or we feel there must be some error in the facts that we have been accepting.

And since such leaps beyond or outside or even apparently against logic sometimes prove to be marvelously insightful, we are unlikely to see the propensity to trust our intuition as a mistake. Indeed our brains plainly work in such a way as to make that simply impossible. At bottom, we have to rely on intuition, and perhaps we have good reason as well as no choice but to rely on intuition. But the proposals of the concluding chapters of this study speak to changes in process that might yield lay risk intuition more in line with expert intuition. So the argument is saying (*a*) that the rest of us would do better (serve our own interests better, or our sense of what we want for our community) if things could be arranged so that we would more often rely on expert intuition rather than our own.

That claim (*a*) has an unfashionable ring today and needs to be made with some care. To start, notice that expertise itself is highly context dependent. No one is an expert in more than a few things, and every expert is one of the rest of us on most matters. So the claim is not that some elite (all-around experts) should have authority over the rest of us (all-around lay people). It is a claim—call it *b*—that since it is hard to question our own intuition (though no reasonable person can doubt that apparently sound intuitions sometimes prove to be wrong), we ought to arrange things in a way that lessens our propensity to be locked into our own intuition even in cases where a clear consensus among people with plenty of experience on a matter sees things the other way. Perhaps that sounds more modest than claim *a*. But I intended no difference between the two claims, and if you find *a* unpalatable, then simply take the claim here to be *b* only. The remedial proposals of chapters 8 and 9 would not be affected in any way.

Nevertheless, even with careful qualification, a claim commending expertise over the intuition of ordinary people readily sounds perverse, and it certainly sounds unfashionable. That is so even though it is a perfectly ordinary intuition—just an appeal to what we all ordinarily would see as common sense—to say that, under odd or unfamiliar circumstances, people may have intuitions that are misleading in terms of their own primary preferences. I may think that I prefer what, if I were better informed, more experienced, or otherwise (by my own standards) in a better position to judge, I would not prefer. And I may feel strongly, or sense as really important what, in a better light, I would see as a mild or inconsequential preference—if, indeed, the preference remains in place at all. And such effects, which no sensible person would deny apply to herself, can only be stronger in the context of social judgment, where a person is commonly dealing with questions out of scale with everyday experience and sees no large consequences to their own judgment (as only a drop in some very large bucket).

Clearly there is a tension between the risks of leaving things to an unchallengeable elite (which much history assures us will be vulnerable to corruption and distortion of perspective) and leaving things to popular sentiment (which much history assures us will be vulnerable to transient passions and misinformation). That popular sentiment is not infallible surely is not a claim that could be seriously controversial, however tactless it might be to say it too bluntly. As Winston Churchill liked to remind people, the merit of democracy is not that it has no flaws, only that it is the worst form of government except for all the others that have been tried.

Social choice is difficult, but the advantage of social cooperation enormous. For some matters, as Adam Smith saw, out of many individual choices, with each considering only things of direct concern to and familiar to the individual, good social choices can emerge, guided as if by a benign, invisible hand. But on many other matters, environmental choices being conspicuous here, social choices that make good sense need to be made socially and cannot be left to emerge from uncoordinated individual choice. We have perverse invisible fist cases, as well as benign invisible hand cases. We need a process that yields one choice for all of us. Sometimes doing that even minimally well will yield large gains over doing without coordinated social choice at all. But social choices can also be terribly bad choices, as in other circumstances uncoordinated market choices can aggregate to terribly bad choices. Making a form of government, including democracy, work well enough to survive is not something that requires no thought and no work.

REFERENCES

Alhakami, A., and P. Slovic. 1994. A psychological study of the inverse relationship between perceived risk and perceived benefit. *Risk Analysis* 14:1085–96.

Baron, J. 1995. *Morality and Rational Choice*. Kluwer.

Breyer, S. 1993. *Breaking the Vicious Circle*. Harvard University Press.

Bruner, J. 1956. *A Study of Thinking*. Wiley.

Carter, L. J. 1987. *Nuclear Imperatives and Public Trust*. Resources for the Future, Johns Hopkins University Press.

Cerf, C., and V. Navasky. 1984. *The Experts Speak*. Pantheon Books.

Coursey, D., et al. 1993. Insurance for low-probability hazards: A bi-model response to risk. *Journal of Risk and Uncertainty* 7:95–116.

Covello, V. 1992. Risk comparisons and risk communication. In *Communicating Risk to the Public*, edited by R. Kasperson and P. Stallen. Kluwer.

Davis, J. A., and T. W. Smith. 1994. *General Social Survey, 1972–1994: Cumulative Codebook*. National Opinion Research Center.

Dawes, R. M. 1979. The robust beauty of improper linear models in decision-making. In *Judgment under Uncertainty: Heuristics and Biases*, edited by D. Kahneman et al. Cambridge University Press.

Dawes, R. M., and B. Corrigan. 1974. Linear models in decision-making. *Psychological Bulletin* 81:95–106.

Douglas, M., and A. Wildavsky. 1982. *Risk and Culture*. University of California Press.

Evans, J. S. B. 1983. *Thinking and Reasoning*. Routledge & Kegan Paul.

Frisby, J. P. 1979. *Seeing: Illusion, Brain, and Mind*. Oxford University Press.

Gardner, H. 1985. *The Mind's New Science*. Basic Books.

Goffman, E. 1959. *The Presentation of the Self in Everyday Life.* Wiley.

Gough, M. 1993. Dioxin: Perception, estimates, and measures. In *Phantom Risk: Scientific Inference and the Law*, edited by K. R. Foster. MIT Press.

Graham, J. D., and J. B. Wiener. 1995. *Risk vs. Risk: Tradeoffs in Protecting Health and the Environment*. Harvard University Press.

Hattchouel, J. M., A. Laplanche, and C. Hill. 1995. Leukemia mortality around French nuclear sites. *British Journal of Cancer* 71 (3): 651.

Hogarth, R. 1987. *Judgment and Choice*. Wiley.

Inhaber, H. 1982. *Energy Risk Assessment*. Gordon & Breach.

Kahneman, D., and A. Tversky. 1984. Choices, value and frames. *American Psychologist* 39:341–50.

Kasperson, R., and P. Stallen. 1991. *Communicating Risks to the Public*. Kluwer.

Kasperson, R., et al. 1987. Radioactive waste and the social amplification of risk.

Kinlen, L. J., M. Dickson, and C. A. Stiller. 1995. Childhood Leukemia near large rural construction cites, with a comparison with Sellafield nuclear site. *British Medical Journal* 310:763-68.

Koehler, B. 1996. The base-rate fallacy revisited. *Behavioral and Brain Sciences* 19, no. 1 (in press).

Kraus, N., T. Malmfors, P. Slovic. 1992. Intuitive toxicology: expert and lay judgments of chemical risks. *Risk Analysis* 12:215–31.

Krimsky, S., and D. Golding, eds. 1992. *Social Theories of Risk*. Praeger.

Krupnick, A. J., and P. R. Portney. 1991. Controlling urban air pollution: A benefit-cost assessment. *Science* 252 (April): 522–28.

Kuhn, T. S. (1962) 1970. *The Structure of Scientific Revolutions*. University of Chicago Press.

Kunreuther, H., et al. 1978. *Disaster Insurance Records: Public Policy Lessons*. Wiley.

Lave, L. 1981. *The Strategy of Social Regulation*. Brookings.

League of Women Voters. 1985. *A Nuclear Waste Primer*. League of Women Voters Education Fund.

Margolis, H. 1977. The politics of auto emissions. *Public Interest.*

———. 1988. *Patterns, Thinking, and Cognition: A Theory of Judgment*. University of Chicago Press.

———. 1991. Free riding vs. cooperation. In *Strategy and Choice*, edited by R. Zeckhauser. MIT Press.

———. 1993. *Paradigms and Barriers: How Habits of Mind Govern Scientific Beliefs*. University of Chicago Press.

Matuszek, M. 1988. Safer than sleeping with your spouse. In *Low-level Radioactive Waste Regulation*, edited by M. E. Burns. Lewis.

Mendeloff, J. M. 1988. *The Dilemma of Toxic Substance Regulation: How Overregulation Causes Underregulation at OSHA*. MIT Press.

NAS (National Academy of Sciences). 1978. Report of the committee for a study on saccharin and food safety policy. NAS.

Neel, J. V. 1994. The problem of "false positive" conclusions in genetic epidemiology: Lessons from the leukemia cluster near the Sellafield nuclear installation. *Genetic Epidemiology* 11:213–33.

Nisbett, R. E., and L. Ross. 1980. *Human Inference: Strategies and Shortcomings of Social Judgment*. Prentice-Hall.

Nisbett, R. E., and T. D. Wilson. 1977. Telling more than we can know: Verbal reports on mental processes. *Psychological Review* 84:231–59.

OTA (Office of Technology Assessment). 1977. *Cancer Testing Technology and Saccharin*. U.S. Congress.

Perkins, D. 1981. *The Mind's Best Work*. Harvard University Press.

Peters, E., and P. Slovic. 1995. The role of affect and worldviews as orienting dispositions in the perception and acceptance of nuclear power. Decision Research Report 95-1.

Rayner, S., and R. Cantor. 1987. How fair is safe enough? *Risk Analysis* 7:3–13.

Reber, A. S. 1992. *Implicit Learning and Tacit Knowledge*. Oxford University Press.

Rodricks, J. 1992. *Calculated Risks*. Cambridge University Press.

Schwarz, M., and M. Thompson. 1990. *Divided We Stand*. University of Pennsylvania Press.

Shweder, R., and J. Haidt. 1993. The future of moral psychology: Truth, intuition and the pluralist way. *Psychological Science* 4 (6): 360.

Slovic, P. 1992. Perception of risk: Reflections on the psychometric paradigm. In *Social Theories of Risk*, edited by S. Krimsky and D. Golding. Praeger.

Slovic, P., B. Fischoff, S. Lichtenstein. 1979. Rating the risks. *Environment* 21:14–20, 36–39.

Slovic, P., J. H. Flynn, and M. Layman. 1991. Perceived risk, trust, and the politics of nuclear waste. *Science* 254:133–38.

Slovic, P., and S. Lichtenstein. 1983. Preference reversals. *American Economic Review* 73:596–605.

Starr, C. 1969. Social benefit vs. technological risk: What is our society willing to pay for safety? *Science* 165:1232–38.

Tversky, A., and D. Kahneman. 1981. The framing of decisions and the psychology of choice. *Science* 211:453–58.

———. 1984a. Choices, values and frames. *American Psychologist* 39: 341–50.

———. 1984b. Rational choice and the framing of decision. In *Rational Choice*, edited by R. M. Hogarth and M. W. Reder. University of Chicago Press.

Vlek, C. A., and P. Stallen. 1980. *Acta Psychologica*, vol. 63.

Viscusi, W. K., ed. 1994. Special issue. *Journal of Risk and Uncertainty* 8 (January).

Wildavsky, A. 1992. *Risk Analysis*. American Academy of Arts and Sciences.

Zajonc, R. 1980. *Emotions, Cognition, and Behavior*. Cambridge University Press.

INDEX